Embedding STEAM in Early Childhood Education and Care

Caroline Cohrssen • Susanne Garvis
Editors

Embedding STEAM in Early Childhood Education and Care

palgrave
macmillan

Editors
Caroline Cohrssen
Faculty of Education
The University of Hong Kong
Hong Kong SAR, China

Susanne Garvis
Department of Education
Swinburne University of Technology
Hawthorn, Australia

ISBN 978-3-030-65623-2   ISBN 978-3-030-65624-9 (eBook)
https://doi.org/10.1007/978-3-030-65624-9

This Palgrave Macmillan imprint is published by the registered company Springer Nature Switzerland AG.
The registered company address is: Gewerbestrasse 11, 6330 Cham, Switzerland

Preface

The inspiration for this book arose a little over one year ago when we were discussing the incorporation of Science, Technology, Engineering and Mathematics (the so-called STEM subjects, and STEAM, when the Arts are included) in early childhood education settings. We realized the extent to which successes and challenges were shared experiences across countries and curricula. In addition, children learn from birth and much of this learning occurs in the home environment. For this reason, this book is aimed at parents and caregivers of children aged from birth to eight years, as well as early childhood professionals. What we did not anticipate was another global experience: COVID-19. What a time it has been.

We would like to thank all the wonderful authors who contributed thought-provoking chapters to this important book on STEAM in early childhood education. We would also like to thank the reviewers for providing such detailed and constructive feedback, again despite challenging work circumstances for many people. We set out to achieve an international flavour and this book is enriched by contributions from specialists from many different countries.

Rationale

Embedding STEAM (Science, Technology, Engineering, Arts and Mathematics) in early childhood education is a hot topic. Enacting a STEAM-based curriculum requires teachers to support children's Science, Technology, Engineering and Mathematics learning within an integrated curriculum that includes the Arts—hence 'STEAM'. Proponents of STEAM argue that it creates opportunities for children to learn creatively, making connections between the five discipline areas. However, many pre-service and in-service early childhood teachers and families are uncertain where to begin.

This book evolved from initial discussions between the editors concerning the absence of clear information for teachers and families about supporting STEAM in the context of early childhood education and care (ECEC). High quality ECEC incorporates both structural components (such as curriculum, resources and teacher preparation) and process components (such as interactions that support learning), and has a strong focus on prioritizing child outcomes within the context of an integrated curriculum. Consequently, we recognized the need for an edited text to demonstrate the 'how' of STEAM in ECEC that would support pre-service and in-service early childhood teachers as well as parents.

We also set out to take academic conversations about STEAM versus STEM to the teachers and families who are enacting STEM and STEAM directly with children. We encourage our readers to reflect on the debate. Indeed, as you read this book, you will observe that chapter authors take differing positions within this debate. Some authors focus on integrating two components of STEAM, some on more than two, some on STEAM and some on STEM. We hope that the book will inspire you and contribute to improved learning outcomes for all children.

Content Organization

As book chapters are inevitably sequential, we have followed the STEAM sequence. The final chapter presents a strong case for STEM (rather than STEAM). We hope this chapter will encourage the reader to reflect upon their STEM versus STEAM philosophy and pedagogy.

In Chap. 1, Cristina Guarrella focuses on science, arguing for a process skills approach and the applicability of process skills across STEAM disciplines. Here, process skills include observing, comparing, classifying, predicting and checking, and communicating and recording. This shifts to focus on science learning from content knowledge to transferable skills.

In Chap. 2, Maria Hatzigianni, Athanasios Gregoriadis, Nektarios Moumoutzis, Marios Christoulakis, and Vasiliki Alexiou provide insights into how a design thinking model that incorporates new technology and the Arts can make abstract concepts such as peace, border disputes, nationalism, and heroes accessible to young children.

In Chap. 3, Suzannie Leung, Kimburley Choi, and Mantak Yuen focus on digital play as they describe how children acquired cinematic language and shared toy-playing stories whilst producing their own one-minute videos.

In Chap. 4, Rhys George and Parian Madanipour discuss the use of technology in children's exploratory and imaginary play with an augmented reality sandbox.

In Chap. 5, Amanda Sullivan and Amanda Strawhacker explore low-cost and hands-on approaches to teaching young children about coding and engineering as they focus on screen-free STEAM.

In Chap. 6, Lyn English highlights the way in which early engineering experiences integrate smoothly into early STEM and STEAM curricula and discusses engineering habits of mind and design processes.

In Chap. 7, Jan Deans and Susan Wright explore how children experience STEAM learning holistically. Drawing on exemplars of the lived experience of three young children, STEAM is presented as an integrated experience for meaning-making.

In Chap. 8, Susan Chapman, Georgina Barton, and Susanne Garvis discuss ways in which an Arts Immersion approach is effective in establishing sustainable teacher confidence and competence in interdisciplinary teaching as well as improving the learning of young children.

In Chap. 9, Marianne Knaus describes ways in which informal, everyday experiences can incorporate opportunities for mathematical thinking and inquiry-based learning whilst children explore their worlds.

In Chap. 10, Karin Franzén applies a mathematics lens to demonstrate toddlers' use of their bodies to explore and understand concepts of shape, location and direction, as well as to problem-solve real-world challenges.

In Chap. 11, Amelia Church and Caroline Cohrssen zoom in closely on the mechanics of interaction by providing insights from conversation analysis, shining a spotlight on real-world mathematics and science discussions with children to show how concept development is supported.

In Chap. 12, Nicola Yelland describes the 'how' of a STEM approach in the early years as children transition from preschool to formal school. The chapter focuses on how collaboration between preschool educators and primary school teachers created learning ecologies to encourage children's agency in their learning.

In Chap. 13, Douglas Clements and Julie Sarama question the inclusion of the Arts in the STEM domain, suggesting that it weakens the subject matter content, as well as the pedagogical and epistemological coherence of STEM. They do not suggest that STEM is more important than other learning domains, but rather that young children need high-quality experiences in them all.

Implications for Initial Teacher Education

Advocating for an interdisciplinary approach has important implications for initial teacher education. Initial teacher education needs to equip teachers with the content knowledge and teaching strategies to facilitate children's engagement with the big ideas of individual discipline areas, *as well as* to equip teachers to enact the 'whole'—an interdisciplinary approach that transcends subject areas. One possible consequence of embedding discipline learning in an interdisciplinary approach is superficial learning: whilst the experience may follow children's interests, plans may not provide optimal opportunities to consolidate and extend learning. In addition, assessment for, and assessment of, learning may consequently be inaccurate. Clearly, an interdisciplinary STEM or STEAM approach requires teachers to possess disciplinary and pedagogical knowledge, and confidence. The priority outcome for teacher education and the providers of in-service professional learning opportunities is thus highlighted: confident teachers who are able to draw on a range of teaching and learning strategies and who are able to balance and support holistic and individual learning across all discipline areas. Given that many teacher education programmes focus heavily on literacy and numeracy at the expense of other learning areas, there is a risk that current graduates will not have the skills needed to be effective STE(A)M teachers, as this requires understanding different ways of thinking and different kinds of 'knowledge'. Examples of these are provided in the chapters of this book.

Another key message across chapters is a focus on the child. Teachers are required to recognize what children know already and what they are ready to learn next to inform child-centred curriculum planning. In the context of STEAM in early childhood education, teachers are required to recognize evidence of Science-, Technology-, Engineering-, or Mathematics thinking as it is communicated by the child in what they say, do, draw, and make—and to respond purposefully to children in order to support the consolidation and extension of conceptual understanding and the language that goes with it whilst incorporating the arts. Intentionality manifests in different ways that are impacted by teacher pedagogical content knowledge, learning trajectories, teaching philosophy, curriculum guidelines, regulations, and national priorities.

Conclusion

The idea for this book was born during a conversation about pedagogy in two countries at opposite ends of the world. Despite their geographical distance, we noted many similar challenges and successes. We hope that this book draws early childhood educators (whether as individuals or as teams) and parents into the conversation. Each chapter of the book adds a new perspective to the conversation and deepens our understanding of how 'to do' STEM or STEAM with young children. We hope it will inspire early childhood educators and families alike to enact STE(A)M teaching with confidence and to facilitate learning by children from toddlers to the early years of school. Finally, we also hope that the book will support conversations about pedagogy and reflective practice.

Hong Kong SAR, China — Caroline Cohrssen
Hawthorn, VIC, Australia — Susanne Garvis

Acknowledgements

We would like to acknowledge the expertise contributed by our panel of specialist reviewers. The contribution of their considered remarks strengthened the book.

Professor Camilla Björklund	*University of Gothenburg, Sweden*
Dr Joanne Blannin	*The University of Melbourne, Australia*
Associate Professor Estelle Blanquet	*University of Bordeaux, France*
Dr Wendy Goff	*Swinburne University of Technology, Australia*
Associate Professor Christine Howitt	*The University of Western Australia, Australia*
Dr Harry Kanasa	*Griffith University, Australia*
Dr Malin Nilsson	*University of Gothenburg, Sweden*
June O'Sullivan, MBE	*Chief Executive, London Early Years Foundation, UK*
Emeritus Professor Bridie Raban	*The University of Melbourne, Australia*
Professor Wee Tiong Seah	*The University of Melbourne, Australia*
Associate Professor Oliver Thiel	*Queen Maud University College, Norway*
Dr Liisa Uusimäki	*University of Gothenburg, Sweden*
Associate Professor Beth van Meeteren	*University of Northern Iowa, USA*
Professor Steve Walsh	*Newcastle University, UK*

Contents

Notes on Contributors

Vasiliki Alexiou has been a kindergarten teacher in Greece for 22 years. Having been awarded a bachelor's and a master's degree in Kindergarten/ Preschool Education, Alexiou's expertise lies in creating curriculum programs for preschool education. For the last 11 years she has been working at the 2nd Model Experimental Kindergarten of Thessaloniki. This kindergarten has close ties with the Aristotle University of Thessaloniki. For the last 7 years, Alexiou has been responsible for the school's English Language Club.

Georgina Barton is Professor of Literacies and Pedagogy in the School of Education, University of Southern Queensland. She is also Associate Head of School—Research. She has experience as Programme Director and has taught English and literacy education and arts education courses in both primary and secondary education programmes. Research areas include literacies, modalities, arts education, and international education. She has utilized a number of research methodologies and methods including ethnography, arts-based research methods, case study, and narrative.

Susan Chapman is a lecturer in the School of Early Childhood and Inclusive Education at the Queensland University of Technology. She has worked as an actor, a musician, and a teacher across education sectors.

Her work on an Arts Immersion includes a research fellowship in STEAM at Griffith University.

Kimburley W. Y. Choi is an associate professor in the School of Creative Media, City University of Hong Kong. Her research interests are wide-ranging and lie at the creative intersection of sociology, anthropology and media studies. She is the author of articles in *Qualitative Research*, *Journal of Consumer Culture*, *Cultural Studies Review*, *Ethnography*, *Urban Studies*, *Journal of Gender Studies*, *Childhood*, *Social Semiotics*, amongst others.

Marios Christoulakis graduated from the School of Electronic and Computer Engineering from the Technical University of Crete in 2008. He holds an MEng from the same school and is a PhD candidate in the School of Architecture of the Technical University of Crete. Also, he is a researcher in the Transformable Intelligent Environments Laboratory. His research interests involve e-learning applications, modern educational technologies, serious games applications, including digital games and digital storytelling employing eShadow, environmental projection applications, and projection mapping technologies.

Amelia Church is a senior lecturer in the Melbourne Graduate School of Education at The University of Melbourne where she teaches subjects in early childhood and applied conversation analysis. Amelia's PhD was published as *Preference Organization and Peer Disputes: How Young Children Resolve Conflict* (2009). More recently, her work has been published in *Children's Knowledge-in-Interaction: Studies in Conversation Analysis* (2016) and early childhood journals. Her research interests include communicative competence in early childhood, peer interaction, and conversation analysis in institutional settings.

Douglas H. Clements is Distinguished University Professor and Kennedy Endowed Chair in Early Childhood at the University of Denver. Focusing on early education, especially mathematics education and educational technology, he has published over 166 refereed research studies, 27 books, 100 chapters, and 300 additional works, and has directed more than 38 funded projects. His contributions have led to the development of new mathematics curricula, software, teaching approaches, teacher training initiatives, and models of 'scaling up' interventions, including the LearningTrajectories.org web application. He has served on the U.S. President's National Mathematics Advisory Panel, the *Common Core State Standards* committee, and the National Research Council's Committee on Early Mathematics, and co-authors each of their reports.

Caroline Cohrssen is an associate professor in the Faculty of Education at The University of Hong Kong (HKU). In Australia, she led the development of influential framework-based curriculum resources in the Northern Territory and Victoria. Since joining HKU, exposure to early childhood teaching and learning priorities in the Asia-Pacific Region, as well as in low- and middle-income countries around the world, is broadening her understanding of early childhood learning and development. Caroline's work addresses the what, why, *and how* questions of STEM. It aims to support parents' and early childhood professionals' ability to recognize children's demonstrations of STEM thinking in what they make, say, and do and to identify next steps for learning that respond to children's interests.

Jan Deans is the Associate Director Early Childhood Education at The University of Melbourne. She is also the Director of the Early Learning Centre, which is the University's research and demonstration preschool. Jan is a long-time advocate for teaching and learning through the arts and has worked both locally and internationally in early childhood, primary, tertiary, and special education settings. She has broadly based expertise in relation to early childhood education and service delivery and her recent research interests include learning through dance, social-emotional competence, and learning in the natural environment.

Lyn D. English is Professor of STEM and Mathematics Education at the Faculty of Education, Queensland University of Technology. She has researched and published over several decades in mathematics education and in STEM more broadly, spanning preschool through year nine. Her areas of research include engineering education and integrated STEM education, mathematical modelling, problem solving and posing, statistics education, mathematical reasoning and development, and early computational thinking and coding.

Karin Franzén is Associate Professor of Early Childhood Education at the School of Education, Culture and Communication at Mälardalens University. Her research interest focuses on toddlers' education in preschool, especially their mathematical learning. Another research interest is educational leadership in schools and preschools. Her research has been published in several international scholarly journals and books. She has a critical research approach and challenges dominant educational discourses. She strives to visualize taken-for-granted normalizations flourishing in educational environments.

Susanne Garvis is Professor of Education (Early Childhood) and the Chair of the Department of Education at Swinburne University of Technology. Her research focuses on policy, quality, and learning within the field of early childhood education and care. She has worked in both Australian and Swedish academic institutions and been involved in a number of national and international research projects within early childhood education. Her research has led to policy development around teacher skills in early childhood education and care.

Rhys George investigated children's use of digital technology in his master's degree research project, reflecting his passion for using innovative technologies as tools for play within social constructivist educational settings. His research uses participatory methods and focuses on early childhood education, augmented reality technology, and learning through play. Enacting theory- and research-informed teaching practice, Rhys is an atelierista at Bold Park Community School in Western Australia. He has presented his research and learning stories around Australia and internationally.

Athanasios Gregoriadis is Associate Professor of Early Years in the Department of Early Childhood Education at the Aristotle University of Thessaloniki. He was the chair of the 29th EECERA Conference 2019, organized in Thessaloniki, Greece. In the past he was a visiting professor at the University of Bielefeld and a visiting research fellow at the University of Oxford. His main research interests include teacher-child relationships, quality of early childhood education environments, professional development, ICTs in preschool centres, and the support of home learning environments. He has published over 50 refereed articles in areas of early childhood.

Cristina Guarrella combines her experience as a three-year-old kindergarten teacher, with pre-service teacher education and research on early childhood science at the Melbourne Graduate School of Education. She strives to make the link between research and practice accessible for educators and is a co-author of the Northern Territory Preschool STEM Games. Cristina's doctoral research is investigating teachers' enactment of science games in metropolitan and regional early childhood settings, with a specific focus on teachers' assessment practices, attitudes and beliefs, and classroom quality. Her ongoing research interests include early childhood and primary STEM education, the role of social media in teacher professional learning, and measuring the impact of STEM interventions on children's learning outcomes.

Maria Hatzigianni is an honorary lecturer at The University of Melbourne. Her expertise builds on a rich early childhood career with more than 12 years of teaching experience in the early childhood sector. Her research focuses on the implementation of digital technologies in education and she has published widely. Her recent research projects include a national project to improve quality for early childhood centres (led by Professor Linda Harrison); the investigation of parents' and teachers' beliefs around the use of touchscreen devices by very young children (birth to three); maker pedagogies, makerspaces, design thinking in the early years of school; and virtual reality in early childhood education.

Marianne Knaus is an associate professor and researcher in the School of Education at Edith Cowan University in Perth, Australia. Marianne coordinates and teaches the early childhood mathematics programmes (birth to eight years) and other units in the undergraduate and postgraduate courses. Her research interests include mathematics, play pedagogy, and the influence of family and community in the transition to school. She has over 30 years' experience as an educator working in a range of early childhood settings in Queensland and New South Wales.

Suzannie K. Y. Leung is an assistant professor in the Department of Curriculum and Instruction, the Chinese University of Hong Kong. She has an interdisciplinary background of visual arts, psychology and early childhood education. Her academic work has explored the possibilities of different forms of visual arts in Hong Kong kindergartens. She is the author of articles in *Young Children*, *Asia-Pacific Journal of Research in Early Childhood Education*, *Journal of Education for Teaching*, *Early Education and Development*, amongst others.

Parian Madanipour is a research fellow at The University of Melbourne where she is project-managing the Melbourne Graduate School of Education's *Early Childhood Professional Practice Partnership Project.* Reflective practice has informed her growing interest in the inclusion of innovative pedagogies and technology in early childhood education and, specifically, the adaptation and promotion of STEAM in early childhood education settings. A qualified early childhood teacher, her master's degree research focused on facilitating young children's spatial thinking through dance.

Nektarios Moumoutzis graduated from the Computer Science Department of the University of Crete in 1992 and holds an MEng in Electronic and Computer Engineering from the Technical University of Crete (1998). In 2002 he joined the Department of Electronic and Computer Engineering (now School of Electrical and Computer Engineering), Technical University of Crete, and since then he belongs to the laboratory teaching staff of the Distributed Multimedia Information Systems and Applications Laboratory. He has been involved in various research projects and his expertise includes project management, design and implementation of modern information systems, object-oriented databases, and e-learning systems.

Julie Sarama is Distinguished University Professor and Kennedy Endowed Chair in Innovative Learning Technologies and at the University of Denver, Colorado, USA. She has taught high school mathematics, computer science, middle school gifted mathematics, and early childhood mathematics. She has directed over 10 projects funded by the National Science Foundation and the Institute of Education Sciences and has authored over 77 refereed articles, 6 books, 55 chapters, and over 80 additional publications. She has also developed and programmed over 50 award-winning educational software products. Her research interests include children's development of mathematical concepts and competencies, implementation and scale-up of educational interventions, professional development models' influence on student learning, and implementation and effects of software environments.

Amanda Strawhacker is a PhD candidate in the Department of Child Study and Human Development at Tufts University's DevTech Research Group, where she works to design, implement, and evaluate novel technologies and curricula to engage young children in topics of the twenty-first century. She has contributed to the research and development of several technologies including the ScratchJr programming app, the KIBO robotics kit, the Early Childhood Makerspace at Tufts, and most recently the CRISPEE bioengineering kit, which she presented at a TEDx talk in 2018 (TEDxYouth@BeaconStreet). CRISPEE is part of a collaboration between DevTech and the Human-Computer Interaction Lab at Wellesley College. Amanda's research focuses on engaging children in playful learning about bioengineering and bioethics.

Amanda Sullivan PhD, is a child development specialist who researches the impact of new technologies on children. Amanda's research focuses on using new technologies to engage girls in STEM (Science, Technology, Engineering, and Mathematics) in order to increase the representation of girls and women in these fields. Her research on gender and technology has been published in numerous academic journals and has been featured in popular outlets such as *WIRED* and *EdWeek*. She is the author of the new book *Breaking the STEM Stereotype: Reaching Girls in Early Childhood* (2019) and co-creator of the *ScratchJr Coding Cards*.

Susan Wright is an honorary professor in the Melbourne Graduate School of Education at The University of Melbourne, where she was formerly the Chair of Arts Education. Previous appointments include the Head of Early Childhood at the National Institute of Education in Singapore and Director of the Centre for Applied Studies in Early Childhood at the Queensland University of Technology. Susan's qualifications, teaching, and research span music-movement and the visual arts. Her research of children's transmediation across modes—embodied, graphic, narrative, visual-spatial, bodily kinaesthetic, musical—surfaces the voices of children through interlocutor-child exchanges. Such discursive interchanges enrich and inform children's conceptual learning and fundamental principles of STEAM, much of which arise from arts-based encounters.

Nicola Yelland is Professor of Early Childhood Studies in the Melbourne Graduate School of Education at The University of Melbourne. Her teaching and research interests have been related to transformative pedagogies and the use of new technologies in school and community contexts. She has also worked in East Asia and examined the culture and curriculum of early childhood settings. Nicola's work engages with educational issues with regard to varying social, economic, and political conditions and thus requires multidisciplinary perspectives.

Mantak Yuen is an associate professor and the director of the Programme for Creativity and Talent Development, Centre for Advancement in Inclusive and Special Education, Faculty of Education, The University of Hong Kong. He is a registered counselling and educational psychologist in Hong Kong. He is the principal investigator of a kindergarten social-emotional learning project. He is the lead editor of the Springer book series *Advancing Inclusive and Special Education in the Asia-Pacific*.

List of Figures

List of Tables

1

Weaving Science Through STEAM: A Process Skill Approach

Cristina Guarrella

Introduction

When people hear 'scientist', they often think of someone conducting experiments in a laboratory. Children learn these ideas from adults in their environment and from representations of science and scientists in books and on television. A well-known study asked children to draw a scientist. Most drawings showed a man in a white coat (Miller, Nolla, Eagly, & Uttal, 2018). A purposeful effort is required to shift this thinking and make science education accessible to all children, regardless of gender or ethnic background. Science is all around us—thus creating the opportunity for every parent, educator and child to be a scientist.

The promotion of children's science process skills begins when a parent or educator notices their emergent skills in play. Recognising what the child can do is the starting point for future learning (Griffin, 2014). Identifying that a child may be comparing the features of different rocks

C. Guarrella (✉)
Melbourne Graduate School of Education, Parkville, VIC, Australia
e-mail: cristina.guarrella@unimelb.edu.au

C. Cohrssen, S. Garvis (eds.), *Embedding STEAM in Early Childhood Education and Care*, https://doi.org/10.1007/978-3-030-65624-9_1

collected in the garden creates a 'teachable moment' (Haug, 2014). The educator has the opportunity to respond to and nurture the development of this skill. They can do so by asking open-ended questions or providing prompts to support engagement of other science process skills. These teachable moments provide a springboard for future learning. By identifying the science process skills children are using in their play, the educator can begin to plan for the future development of these skills.

Many early childhood educators and parents are eager to talk about our natural world and other topics that relate to science, but research has shown that they often feel nervous about doing so and, as a result, may avoid it (O'Brien & Herbert, 2015; Pendergast, Lieberman-Betz, & Vail, 2017). Early childhood educators express concern that they might not be able to answer children's questions because they lack confidence in their own science knowledge. Consequently, early childhood educators may provide children with interesting materials to explore independently (Tu, 2006). Without adult interaction, opportunities for scientific thinking and reasoning, and back-and-forth conversations are limited.

Sometimes, early childhood educators and parents focus more on the end result of a science learning experience than on the learning process. For example, when toddlers are blowing soap bubbles, the focus may be more on the bubbles (the product) than on how the bubbles were made (the process). The dichotomy of process versus product is widely discussed in visual arts education (McLennan, 2010). At the simplest level, the distinction of process and product can be applied to science. Skills such as observing, classifying, comparing, predicting and checking, and communicating and recording are the processes of science. Scientific content knowledge is the product of science. Whilst the focus of this chapter is on the process skills of science, scientific knowledge is still important. Both scientific content and process skills are required for science learning (Zimmerman, 2000). Many early childhood learning frameworks do not prescribe specific science content areas (e.g. the Early Years Learning Framework for Australia (Department of Education Employment and Workplace Relations, 2009)). This is because following children's interest in the world around them (known as an 'emergent curriculum') leads to an infinite number of science content areas to be explored.

Science Process Skills in Early Childhood

In this section, the science process skills typically observed in early childhood are presented individually (Table 1.1). However, these behaviours are woven through children's play and hence the skills are inter-connected. Mathematics learning trajectories are well established (Clements & Sarama, 2014). For example, we know that children need to know number words before they can tag objects with number words. However, specific science process skill trajectories have not yet been identified (Jirout & Zimmerman, 2015). For this reason, the skills that the children are already using in their play show early childhood educators and parents what they already know and what children are ready to learn next.

Observing

Observation is at the core of scientific discovery. Jane Goodall made detailed observations of chimpanzees in Tanzania and discovered chimps use twigs as tools, have their own personalities and were more like humans than anyone had previously thought. Gregor Mendel noticed the variation in the pea flowers growing under the window of his room and conducted a series of experiments breeding peas, leading to the first understanding of the inheritance of traits. To this day, scientists continue to make new discoveries by observing and noticing changes. Observing is a foundational science process skill.

Table 1.1 Science process skills in early childhood

Observing	Using senses to notice the details and changes in living and non-living things in the natural and built environment
Comparing	Identifying similarities and differences in living and non-living things
Classifying	Grouping living and non-living things based on their features and properties
Predicting and checking	Using information gathered from observations to make simple statements about what will happen and conducting simple experiments to check if it happens
Communicating and recording	Using representation to record, organise and communicate scientific ideas, discoveries and understandings

Observation relies on a range of senses: touch, sight, taste, smell, hearing and feeling. Observation is a skill that children use spontaneously with little encouragement. A young child learning to eat solid food will observe how the food feels between their fingers, respond to how it smells and tastes, and even look closely at the food in their hand. In early childhood science education, observing involves the use of senses to notice the details and changes in living and non-living things in the natural and built environment. An observation could take the form of a child hearing a bird call and turning their head to see where the sound came from. With encouragement from an adult, the child may move closer to the bird to have a better look, try to make their own bird call or listen carefully for other birds in the area. In these scenarios, the child's interest in birds is creating a scientific context in which they practise the skill of observation in its simplest form.

As observation skills develop, a child may begin to notice how objects change over time. For example, the child may notice a plant growing or food scraps decomposing in the compost bin. When educators encourage children to return to their observations at different times of the day, or at the same time over a number of days, they 'scaffold' this observation skill. A scaffold is a supportive framework provided by an educator to guide and assist children's learning (Wood, Bruner, & Ross, 1976). Incorporating scientific tools such as magnifying glasses, binoculars or digital microscopes allows children to observe objects and living things on a new scale and extend the role of a scientist in their play.

Incorporating other science process skills such as 'recording' by using drawings or taking photographs of observations, will provide the platform for children to 'compare' observations—to rehearse another science process skill. This example demonstrates how the process skills of observing, recording and comparing can be purposefully supported whilst following the child's interests.

Comparing

Scientists are on a continuous quest to discover new information. To distinguish between 'new' and 'known', scientists make comparisons between their discoveries and the discoveries of those who came before them. As a result, comparison is a cornerstone of scientific exploration. Further, comparisons are frequently made during scientific inquiry as scientists compare different volumes of liquids in chemistry, distances in physics and cell composition in biology.

Children demonstrate scientific comparison in their play when they identify similarities and differences in shells collected from the beach, insects found beneath a rock or the textures of different fabrics. Comparison builds upon observation skills as children discuss similarities and differences in what they see, hear, feel, smell and taste. This can lead to the classification of living and non-living things. Here, the integrated nature of the science process skills is evident as the development of one skill, in turn, supports the development of another. When children begin to test their ideas, comparison supports children to explore what happens when they change one or more of the conditions of their test, as demonstrated in the example below.

Example 1: Comparing in Play

Audrey is playing with cars on the ramp near the preschool room. She is pushing each car and chasing it down the ramp, looking closely at where the car stops. The educator notices this and begins a conversation with Audrey, who says she wants to make the cars drive further at the bottom of the ramp. The educator prompts her to make connections to past experiences of playing with cars. In their discussion, Audrey talks about playing cars at her grandparent's house. She recalls that the cars go really far with one push on the kitchen floor but they don't move at all in the lounge room on the carpet. This provides the educator with the opportunity to support Audrey to investigate the movement of cars from her prior knowledge. Together the educator and Audrey gather materials to test a range of surfaces on the ramp. They roll the car down each surface and measure how far the car travels from the bottom of the ramp. By making comparisons between these distances, Audrey is able to answer her initial question about how to make a car travel further: by changing the surface.

Classifying

Classification systems enable scientists to make sense of large amounts of information. Classification helps scientists to make predictions about 'gaps' in information. These are areas for future research. Some scientific classification systems, such as the magnitude system that classifies the brightness of stars, have been traced back to the times of ancient Greece. New systems continue to be created and evolve based on modern discoveries. For example, one system for biological classification, the Linnaean Taxonomy, was created in the 1700s. This system is continuously developing as modern scientific discoveries, such as DNA technology, help to increase our understanding of the origins of animals and plants.

For young children, classification involves the exploration of features and properties of living and non-living things in order to group them. Spontaneous classifying that occurs in early childhood settings is governed by a child's scientific interest: they may classify leaves, rocks or a range of materials from the built environment. Many young children are fascinated by dinosaurs. It is not uncommon to find four- or five-year-old children debating the features of a velociraptor and a tyrannosaurus rex and then taking on the key characteristics of these animals in their play. Pointing out that dinosaurs with large jaws were meat-eaters and dinosaurs with small jaws ate plants helps children to classify dinosaurs independently and enriches their play.

Teachers or parents who are attuned to a child's scientific classifications point out additional features of the object in which the child is interested. For example, if a child is collecting rocks and observing them through a magnifying glass, an educator who asks open-ended questions about the rocks' features scaffolds the child's classification skills. Hints to encourage the child to classify objects based on size, colour, uses, parts or shapes extend the child's thinking. As a next step, prompting the child to make groups based on these features will extend the development of this skill. Here the child is using their knowledge of the features to classify the objects into set groups. As the children practise this skill, encourage them to identify and describe their own rules for grouping.

Predicting and Checking

Predicting and checking describes children making predictions and seeking information that confirms or changes what they know. The process of looking for information that supports evidence-based decisions is an important science process skill. It is also an important life skill to learn.

The starting point for predicting and checking is simple trial and error. Predicting requires children to use information gathered from previous observations to develop statements about what they think will happen. As children experiment to check their predictions, they develop new knowledge. The new knowledge may confirm the prediction or identify that something else is happening. This provides the child with an opportunity to modify their prediction based on the new observation and to continue with their exploration. Providing children with multiple opportunities to practice predicting and checking allows them to consolidate the knowledge that they are developing through this process.

Everyday events provide opportunities to practise predicting and checking with young children. Taking a moment to look out of a window and ask, 'What do you predict the weather will be like today?' sets up a realistic and meaningful opportunity for children to predict and check. Cooking experiences allow children to explore the scientific concept of physical change (e.g. melting or freezing). Combined with intentional questioning from the educator, stopping at key points to ask children what they think will happen next and encouraging them to discuss what they saw happen, allows for additional process skill development.

Communicating and Recording

Scientists contribute to society by disseminating their research at conferences, in books and research articles, on television and through social media. New knowledge informs industry, business, medicine and a range of other disciplines. Young children should be encouraged to communicate their knowledge as well. Communication can start as simply as a verbal or non-verbal response to something a child has observed in the immediate environment and develop into explanations and justifications for new understandings. For example, an infant may smile or giggle in response to feeling a new texture with their hands, a three-year-old may describe a texture as soft, and a six-year-old may explain a feather is soft so a bird can protect its chicks. When children communicate their knowledge, educators and parents gain insight into the child's capabilities. For educators, this is the starting point to plan learning experiences that consolidate and extend thinking.

'Recording' provides a concrete record of the scientific exploration and supports the development of almost all science process skills. Recording *observations and outcomes of experiments* equips children to compare, discuss and explain scientific ideas and concepts. It enables children to return to their ideas again and again and to make comparisons. Recording *predictions* supports the checking process and equips children to reconsider their ideas when predictions and findings do not match.

The process of recording offers children an opportunity to reflect on their own learning and to use scientific language to discuss their findings. These processes both depend on and encourage higher-order thinking.

In this section, we have described science process skills in early childhood. Using examples, we have identified why these skills are important and ways to support children's development of each skill. Table 1.2 sets out a summary of the what, why and how of science process skills in early childhood for easy reference.

Implementing a Science Process Skills Approach

Before implementing a science process skills approach, educators need to understand the context in which the scientific learning is occurring. Working with your centre's educational leader to answer reflective questions such as those below may be helpful. Understanding the perspectives of the key players in the child's immediate learning environment provides a platform for identifying the child's current science knowledge. It is also

Table 1.2 The what, why and how of science process skills in early childhood

What is the science process skill?	Why is this science process skill important?	How can I support children's development of this skill?
Observing Using senses to notice the details and changes in living and non-living things in the natural and built environment	Foundational skill at the core of scientific inquiry Facilitates children's exploration of the world around them, generating curiosity	Provide the names of living and non-living things as children look at them Ask children open-ended questions about what they observe, for example, what can you see? What does it smell like? What does it feel like? Provide opportunities for children to observe changes over time
Comparing Identifying similarities and differences in living and non-living things	Comparisons allow for observation of change Facilitates future classification of living and non-living things	Support children to describe the features of objects they are observing Identify objects that look the same and different Use descriptive language like more, less, longer, shorter, faster, slower, heavy, light, rough and smooth

(*continued*)

Table 1.2 (continued)

What is the science process skill?	Why is this science process skill important?	How can I support children's development of this skill?
Classifying Grouping living and non-living things based on their features and properties	Develops systems for ordering Supports understanding of relationships between groups of living things	Order a variety of objects into groups based on size, shape, colour, uses, materials or parts Ask the children open-ended questions like these: 'Why have you grouped these objects together?' 'How did you know which object to put in each group?' Provide opportunities for children to classify objects based on rules that cannot be changed and rules that they create themselves
Predicting and checking Using information gathered from observations to make simple statements about what will happen and conducting simple experiments to check if it happens	Creates opportunities for children to develop new knowledge as they either confirm their predictions or identify that something else is happening	Ask children to make predictions in everyday scenarios, for example, 'What do you predict the weather will be like today?' Record children's predictions so they can revisit them once they have checked them Ask the children, 'Why do you think that happened?' and 'Was your prediction correct?'

(*continued*)

Table 1.2 (continued)

What is the science process skill?	Why is this science process skill important?	How can I support children's development of this skill?
Communicating and recording Using representation to record, organise and communicate scientific ideas, discoveries and understandings	Reflection on thinking and communicating these ideas promotes higher-order thinking Recording scaffolds the development of other science process skills	Support the children to explain their reasoning Encourage the children to record their findings by drawing or taking photographs

necessary to understand factors that may impact how you set about implementing a science process skills approach.

- What are the children's science interests?
- What are the parents' attitudes towards science education?
- How is science education positioned within your early childhood centre philosophy?
- How do educators feel about supporting science learning?

The answers to these questions will influence the way a process skills approach to early childhood science education is enacted.

Assessment for Learning

Planning for learning that authentically follows a child's interests must start with what a child already knows. Assessment for learning is the process of collecting evidence of what a child can do, say, make or draw in order to make informed decisions on how to support their future learning (Griffin, 2014).

In the context of a science process skills approach, this involves educators being sensitive to the scientific process skills that children demonstrate in their play, and recording these in the form of child observations. This requires the educator to be alert to a child who observes, compares, classifies, predicts and checks, and communicates and records. You may choose to prioritise one specific scientific process skill.

Purposeful observations require the educator to be purposeful in their 'looking'—to have a purpose for the child observation that is decided ahead of time. Often, when documenting the observed process skill, the context in which it occurs will also provide an indication of the child's emerging science interests, so it is important to document this also. For example, an educator may observe a child comparing features of bugs (also called minibeasts and scientifically known as invertebrates) living beneath a log. The science process skill the educator would record is 'comparing'. The context in which the behaviour occurred was living things, bugs and minibeasts. If this is not immediately clear, further observations of the child's science interests may be required. High-quality observations of children's learning require appropriate, adequate, accurate *and* authentic observation evidence (Griffin, 2014). Table 1.3 includes prompting questions to help you reflect whether you are gathering the type of meaningful information to which Griffin (2014).

Once appropriate, adequate, accurate and authentic assessment evidence has been collected, the aim is to identify the child's current science learning and consider their future learning. This requires the teacher to analyse the observation. There are several steps in this process.

1. Identify the process skill used by the child. Although the observation started with a clear purpose, it is possible that different or additional science process skills were observed.
2. Describe how the child demonstrated the scientific process skill. For example, the child demonstrated the process skill of observation by

Table 1.3 Examples of prompting questions to support meaningful observations

Appropriate	Does the evidence help me identify the child's use of science process skill(s)?
Adequate	Do I have enough evidence? Will I need to take more observations?
Accurate	Is the observation free of judgement? Is the observation evidence a record of what I see and hear, not what I think?
Authentic	Is this observation evidence for the child I am observing?

using a magnifying glass to look closely at patterns in the petals of a flower.

3. Decide whether you intend to consolidate or extend an observed skill, or to introduce another skill. Here, it is recommended that you refer to documents outlining the development of process skills (such as this chapter), research articles and taxonomies of learning to inform your decision. Further, comparing multiple observations of the child's science process skills will help you to make this assessment.

After analysing the child observation, the next step is to plan learning experiences that provide opportunities for the child to rehearse or extend the process skill, or to explore a different process skill. Your plan will be informed by the analysis of your observation. Often, your observations of one child or a small group of children will inform planning for the whole group of children.

Start by defining learning objectives based on the science process skill to be developed. These take the form of short statements. For example, if you assessed that a child was ready to conduct simple experiments to test whether their predictions were true, the learning objectives could read as follows:

For the child(ren) to:

1. Make predictions based on their previous knowledge or observations
2. Test their predictions
3. Suggest reasons for what they observed

These objectives are specific to the science process skill but can be applied to a number of learning experiences. Transferrable learning objectives based on the science process skill, rather than the content being delivered, ensure that science learning is not lost when the direction of play changes. This highlights the importance of having clear learning objectives: they allow educators to be flexible and to adapt their teaching to follow children's interests without abandoning the learning objectives.

The learning objectives, which have been informed by child observations, provide the structure of play-based learning. Once the objectives are defined, planning ways to scaffold and promote these skills in play can begin. This is where educators demonstrate boundless creativity!

Through rich dialogues with children, parents and educators can promote children's higher-order thinking. Using open-ended questions and asking the children to explain their thinking allows them to reflect, evaluate and justify their new understandings (Wasik & Hindman, 2014).

Integrating Science Process Skills with Technology, Engineering, Arts and Mathematics

Multiple opportunities for language, literacy, arts and numeracy are also promoted when taking a process skills approach to science. When teachers are alert to children's demonstrations of science process skills across the breadth of an informal, play-based curriculum, opportunities to support process skills quickly become apparent. Responding to such teachable moments supports an integrated approach to STEAM education. Figure 1.1 highlights some examples of how science process skills emerge in mathematics, arts, engineering and technology in early childhood.

Reflection

Introducing a science process skills approach into a play-based curriculum may sound like a daunting task. However, the focus on process skills, rather than content knowledge, changes the focus from the educator needing to have all the answers to allowing educators and parents to respond to children's scientific curiosity.

Reflective practice is central to early childhood education and care (Marbina, Church, & Tayler, 2010). Here, a case study is presented that demonstrates how one early childhood teacher reflected on their transition to a science process skills approach in a kindergarten programme attended by children aged three to four years.

The case study demonstrates the challenges and successes achieved by the teacher, highlighting obstacles and explaining how obstacles were overcome to allow for further integration of science into the emergent curriculum approach followed in the classroom.

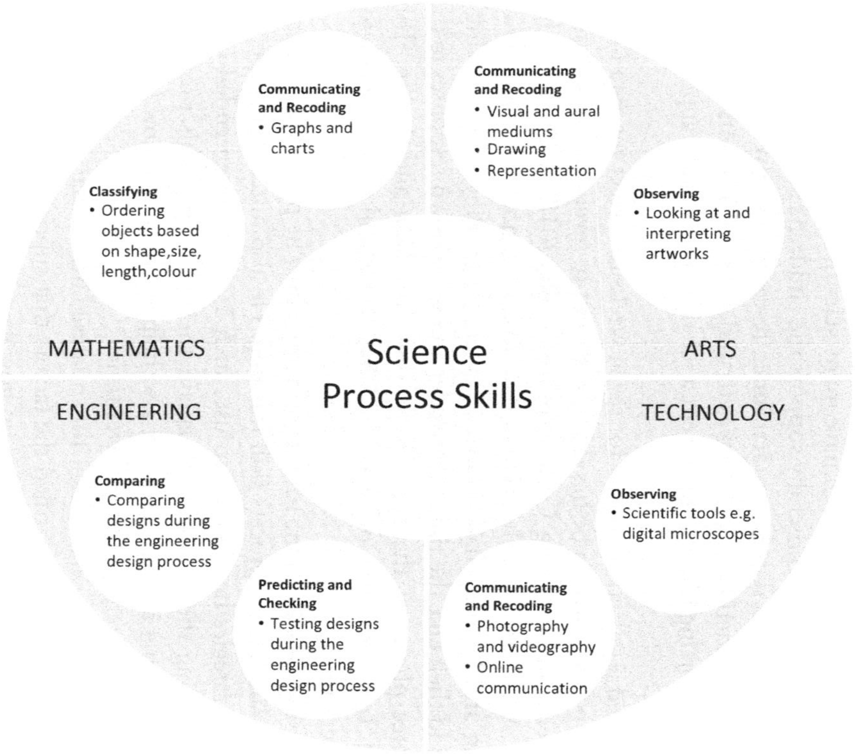

Fig. 1.1 Examples of science process skills in technology, engineering, arts and mathematics

Case Study: Enacting a Science Process Skills Approach Through Reflective Practice

Tom teaches in a three-year-old kindergarten programme located in an inner-city suburb of Melbourne, Australia. Tom believes that science is an important learning area to capture in an emergent curriculum. He noticed that while the children have many scientific interests, he was not finding the time to purposefully support the children's science learning.

Tom decided to conduct practitioner research, by investigating his own teaching practice. His focus was to understand the factors challenging and supporting him to implement a science process skills approach in his teaching.

Tom used a reflective journal for seven weeks to collect his data (Tom was sick in week two, so he has data for six out of seven weeks). He used the Experiential Learning Cycle (Kolb & Kolb, 2018) to structure his journal. At the end of each teaching day, Tom took notes of what science teaching and learning had occurred under the heading 'Experiencing'. He would then take the time to consider and write down responses to the next three phases of the cycle; Reflecting, Thinking and Acting. This allowed Tom to reflect on what had happened and make a plan for future action.

Alongside his reflection, Tom documented how he planned to support the children's science process skill learning. He also noted down specific teaching strategies he planned to use, such as asking open-ended questions or asking the children to explain their thinking. This prompted both his teaching and reflection. It also helped him to see how his plans and teaching strategies changed over time.

At the end of the seven weeks, Tom analysed his journal. He looked for common themes and identified the factors that supported or challenged him to teach science. He counted up how many times each of the factors occurred during the seven-week period. Tom identified that planning and preparing resources (that created the opportunity to practise the skill of observation) most frequently supported him to implement a science process skill approach. The biggest challenge for Tom was when his

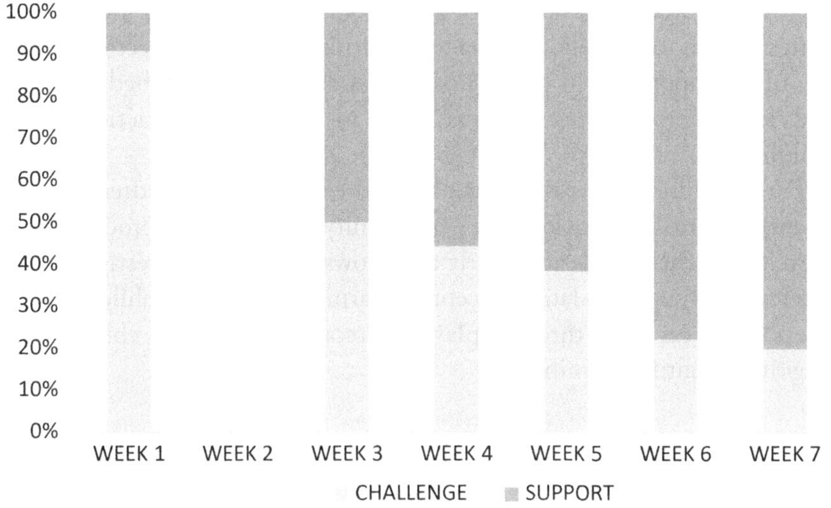

Fig. 1.2 Percentage of challenging and supportive factors over seven weeks of reflective practice

priorities were directed towards tasks unrelated to science or the children's other learning goals.

Tom also looked at the data he collected to identify how his reflective practice influenced the presence of the challenging and supportive factors over the seven weeks. He found that the number of challenging factors decreased and supportive factors increased, as shown in (Fig. 1.2).

Through reflective practice, Tom was able to adopt effective pedagogies to incorporate the use of science process skills into everyday interactions with the children.

Conclusion

In this chapter, we have explored a process skills approach to learning science in early childhood. Developing the process skills of observing, comparing, classifying, predicting and checking, and communicating and recording allows children to expand their scientific curiosity in partnership with their parents and educators. Identifying the science process skills in children's play allows for purposeful and responsive interactions with children to consolidate or extend learning. Further, reflecting on the use of this approach in practice supports educators to embed teaching and learning of science process skills in everyday interactions with children.

Early childhood is a time for play and exploration. Children's innate scientific curiosity provides the opportunity to foster key process skills. Developing these skills at an early age allows children to investigate their interests and supports later conceptual learning. Equipping children with science process skills through playful interactions prepares children for ongoing scientific learning.

References

Clements, D. H., & Sarama, J. (2014). *Learning and teaching early math: The learning trajectories approach* (2nd ed.). New York: Routledge.

Department of Education Employment and Workplace Relations. (2009). *Belonging, being & becoming: The early years learning framework for Australia*. Barton, ACT: Commonwealth of Australia. https://doi.org/10.1037/e672772010-001

Griffin, P. (Ed.). (2014). *Assessment for teaching*. Port Melbourne, VIC: Cambridge University Press.

Haug, B. S. (2014). Inquiry-based science: Turning teachable moments into learnable moments. *Journal of Science Teacher Education, 25*(1), 79–96. https://doi.org/10.1007/s10972-013-9375-7

Jirout, J., & Zimmerman, C. (2015). Development of science process skills in the early childhood years. In *Research in early childhood science education* (pp. 143–165). Dordrecht, The Netherlands: Springer Netherlands. https://doi.org/10.1007/978-94-017-9505-0_7

Kolb, A., & Kolb, D. (2018). Eight important things to know about the experiential learning cycle. *Australian Educational Leader, 40*(3), 8–14.

Marbina, L., Church, A., & Tayler, C. (2010). *Practice principle 8: Reflective practice*. Melbourne, VIC: Department of Education and Early Childhood Development.

McLennan, D. M. P. (2010). Process or product? The argument for aesthetic exploration in the early years. *Early Childhood Education Journal, 38*(2), 81–85. https://doi.org/10.1007/s10643-010-0411-3

Miller, D. I., Nolla, K. M., Eagly, A. H., & Uttal, D. H. (2018). The development of Children's gender-science stereotypes: A meta-analysis of 5 decades of U.S. draw-a-scientist studies. *Child Development, 89*(6), 1943–1955. https://doi.org/10.1111/cdev.13039

O'Brien, F., & Herbert, S. (2015). Colour, magnets and photosynthesis. *Australasian Journal of Early Childhood, 40*(01), 42–26.

Pendergast, E., Lieberman-Betz, R. G., & Vail, C. O. (2017). Attitudes and beliefs of prekindergarten teachers toward teaching science to young children. *Early Childhood Education Journal, 45*(1), 43–52. https://doi.org/10.1007/s10643-015-0761-y

Tu, T. (2006). Preschool science environment: What is available in a preschool classroom? *Early Childhood Education Journal, 33*(4), 245–251. https://doi.org/10.1007/s10643-005-0049-8

Wasik, B. A., & Hindman, A. H. (2014). Realizing the promise of open-ended questions. *The Reading Teacher, 67*(4), 302–311.

Wood, D., Bruner, J. S., & Ross, G. (1976). The role of tutoring in problem solving. *Journal of Child Psychology and Psychiatry, 17*, 89–100.

Zimmerman, C. (2000). The development of scientific reasoning skills. *Developmental Review, 20*(1), 99–149. https://doi.org/10.1006/DREV.1999.0497

2

Integrating Design Thinking, Digital Technologies and the Arts to Explore Peace, War and Social Justice Concepts with Young Children

Maria Hatzigianni, Athanasios Gregoriadis, Nektarios Moumoutzis, Marios Christoulakis, and Vasiliki Alexiou

Introduction

Early childhood education advocates for holistic and integrated approaches which are interdisciplinary and reflective. Learning is based on real-world experiences, inquiry and experimentation. Early childhood education does not support the teaching of separate subjects (e.g., science,

---

M. Hatzigianni (✉)
Melbourne Graduate School of Education, The University of Melbourne, Melbourne, VIC, Australia
e-mail: maria.hatzigianni@unimelb.edu.au

A. Gregoriadis • V. Alexiou
Department of Early Childhood Education, Aristotle University of Thessaloniki, Thessaloniki, Greece
e-mail: asis@nured.auth.gr; vasalexiou@sch.gr

C. Cohrssen, S. Garvis (eds.), *Embedding STEAM in Early Childhood Education and Care*, https://doi.org/10.1007/978-3-030-65624-9_2

maths etc.), instead engages children with topics or themes or projects to promote the twenty-first century competencies (Crockett, 2011; Honey, Pearson, & Schweingruber, 2014). In alignment with this view of education, integration of Science, Technology, Engineering, Arts and Mathematics (STEAM) as a new trend can be easily adopted and followed (Bevan, 2017; Kermani & Aldemir, 2015; Moomaw & Davis, 2010).

The STEAM approach to learning helps children integrate knowledge across disciplines and develop thinking skills in a more systemic and connected way. This approach is often combined with design thinking skills (Stevenson, Bower, Falloon, Forbes, & Hatzigianni, 2019). These are thinking skills that pertain to the design process, a concept that has solid traction in research both past (Buchanan, 1992) and present (Filatro, Cavalcanti, & Muckenberger, 2017). Often relating to approaches to design in industry, educators may draw on principles, cases and strategies for supporting the design process. As an instructional model, 'Design Thinking' (for clarity, title case) is widely used in education. Several models exist, each with slightly different emphases and design stages (Bower, Stevenson, Falloon, Forbes, & Hatzigianni, 2018). Common themes that run through the models include exploration and interpretation in the early stages, generation of ideas in the mid-stages, and testing, evaluating and evolving in the latter stages.

In this study we adopted the IDEO model which encompasses five stages: (a) discovery, (b) interpretation, (c) ideation, (d) experimentation and (f) evolution. Among the available models, the IDEO model (Fierst, Diefenthaler, & Diefenthaler, 2011) is particularly helpful in providing teachers with a detailed free handbook that includes scaffolds, stimuli, first-hand accounts and expert advice.

Studies on the design process for children have also established connections between design thinking skills, and creative and critical

---

N. Moumoutzis • M. Christoulakis
School of Electrical and Computer Engineering, Technical University of Crete, Crete, Greece
e-mail: nektar@ced.tuc.gr; christoulakis@ced.tuc.gr

thinking skills (Shively & Stith, 2018) arguing for the importance of those skills for future generations. However, limited research exists in the use of design thinking with preschool children aged from four to six years (see, e.g., Hatzigianni, Stevenson, Falloon, Bower, & Forbes, 2020). This study addresses this gap, not only using design thinking and working with young children, but also using the design approach in conjunction with the arts to solve conceptual—rather than simply practical—problems. Previous studies exploring the potential of the arts in clarifying abstract and complex concepts have been attempted (see, e.g., Hatzigianni, Miller, & Quiñones, 2016), but design thinking approaches have not been used as widely to address theoretical or conceptual challenges.

The Educational and Social Context of the Study

Early Childhood Education and Curriculum in Greece

In Greece, preschool education is provided in kindergartens ('nipiagogia' translated is a place in which to care for and educate young children, 'nipia'). Kindergartens operate independently or with state primary schools for children aged four to six years old. Since September 2007, under the provisions of Law 3518/2006, the second year of kindergarten is compulsory. Preschool teachers who work in the public kindergartens are highly trained, having completed a four-year bachelor degree in early childhood education at an Hellenic University department of education (Sofou & Tsafos, 2010).

Kindergartens operate under the guidance of a detailed, national curriculum for early childhood education which was first introduced in 2002 (Cross-thematic Curriculum Framework Syllabus Design, [CTC], 2002). The Institute of Educational Policies (IEP), an organisation supervised by the Hellenic Ministry of Education and Religious Affairs, created and facilitated the implementation of this preschool curriculum across the country. CTC has a focus on child-led and child-directed pedagogies which also emphasise the exploration of children's interests and teamwork. The teacher's role is to coordinate, facilitate and scaffold children's learning. One of the main characteristics of the CTC is the cross-thematic approach to learning where the different areas (e.g., music, arts, science) overlap and complement each other (Sofou & Tsafos, 2010).

The Macedonian Crisis

The 'Macedonian conflict or struggle' was a series of social, political, cultural and military disputes that were mainly fought between Greek and Bulgarian subjects who lived in Ottoman Macedonia between 1893 and 1908 (Koukoudakis, 2018; Zahariadis, 1994). The conflict was part of a wider rebel war in which voluntary rebel groups of Greeks fought against Bulgarians and Serbs. Greeks had concerns about their neighbours' intentions and tried to interrupt their expansion plans but also to save Greek schools, religion and language (Zahariadis, 1994).

In the late 1980s, Greece initially opposed the breakup of the Yugoslavian Federation and recognition of its constituent republics as independent states. However, Greece was alone in this battle, and after a series of failed negotiations to end the hostilities, Greece gave its consent together with other European Community members and the US, and recognised Croatia, Slovenia and later Bosnia-Herzegovina. Greece, however, remained opposed to the recognition of the Former Republic of Macedonia and secured the European Community's commitment in 1992 that the former republic would not be recognised until it relinquishes the term 'Macedonia'. Greece argued that this name would be the first step for raising claims of territorial ambitions in the future. Several agreements and disagreements have taken place over the last 40 years which are beyond the scope of this chapter to analyse. Nevertheless, the crisis was never resolved, and this has led to numerous disputes, tensions and protests in Greece (Exarchate et al., 2019; Koukoudakis, 2018; Zahariadis, 1994).

The Prespes Agreement and Public Reactions

The Prespes Agreement reached on 12 June 2018 between Greece and North Macedonia, under the auspices of the United Nations, resolved a long-standing dispute over the latter's name. Signed beside Lake Prespa from which it took its name and ratified by the parliaments of both countries on 25 January 2019, it came into force on 12 February 2019 when the two countries notified the United Nations of the deal's completion.

The Educational and Social Challenge Informing This Study

In Greece, the Prespes Agreement had direct negative political implications. There were large public demonstrations in 2018 and 2019 against the Prespes Agreement which lasted for several days. Student protests took place. The majority of people all over the country were dissatisfied with the result and the name given, and this was particularly evident in the northern part of Greece, in Macedonia, where this study took place. Young children, as active citizens and contributors to their social world, were worried and uncertain about what was happening all around them. Abstract concepts, such as war, peace, border disputes, social justice, nationalism and others, were frequently brought up by media and adults in their lives. Teachers were trying to explain these difficult concepts to children but at the same time tried to be as fair and optimistic as possible, emphasising that a war between the two countries would not solve any problems. In this attempt to unpack the meaning of real-life complex political issues, teachers adopted a design thinking approach to frame their activities and projects. Together with arts and digital technologies, they were able to explore and deepen children's knowledge and understanding of the issues.

This chapter is structured according to the IDEO design thinking model adopted for this case study. Each stage of the IDEO model is briefly explained at the start. The practical implementation of each stage is then described and finally concluding thoughts are outlined. Following the IDEO structure contributes to exploring the model in depth and also clarifies how the different stages of the model were implemented (the 'what', the 'why' and the 'how'). Data were collected over a period of eight months, from November 2018 to June 2019. Prior to data collection, the Research Ethics committee of the Aristotle University of Thessaloniki approved the ethics of the research design in order for researchers to obtain access to the participating kindergarten schools. The school counsellors and the directors of the kindergartens were also informed about the purpose and the methodology of the study. All kindergarten teachers were informed about the study and consented to participate in it. The parents were informed in detail and signed consent forms were obtained. Finally, children were informed about the purpose and the content of the study and advised that they could stop participating in the study at any time. Confidentiality and anonymity were ensured, and pseudonyms are used in this chapter.

The Design Thinking Stages

Four public kindergartens (four- to six-year-old children) from the prefectures of Thessaloniki and Imathia in Northern Greece (Macedonia region) participated in this study. Two of the kindergartens were situated in an urban area and two were situated in a rural area. Seven female kindergarten teachers with a mean age of 46 years and a mean of 16.5 teaching years, and 76 kindergarten children and their parents participated in the study (41 girls and 35 boys). The community (e.g., administrators; museums) around each kindergarten was also informed about the aims of the study.

Discovery

This was the initial stage of the model during which all the preparation and organisation took place and all participants built a clear understanding of what the problem was, what the challenges were and how meaningful solutions could be reached. During this stage, children, parents, teachers and the community were all informed and their feedback was taken into consideration. In the discovery stage, deep engagement and dialogues were fundamental for increasing the chances of successfully addressing the challenges, but also for catering to children's needs and interests.

In this first phase, teachers explored children's existing funds of knowledge about the constructs of war, border disputes, justice and especially their knowledge of the period of the Macedonian struggle. To achieve this, teachers conducted semi-structured interviews with children. They asked questions around children's interpretation of war, peace, Macedonia, the crisis, the struggle, heroes and other related topics.

War and peace were concepts with which most children were familiar. Children could provide some concrete examples of what they understood about these concepts. For example, Irene (four years old) said that *'war is when we fight and kill the bad guys'*. Sunny (four years old) explained: *'war happens when they try to take over Greece'*. A lot of children referred to specific wars Greece was involved in and words like *'fighting'*, *'guns'*, *'killings'*, *'hurting'*, *'shooting'* and *'bad'* were often reported in their interviews. A negative view of war was apparent in all children's answers; whereas peace was seen as positive.

In contrast, children found it difficult to explain *'Macedonia'*, *'crisis'* and *'struggle'*. A few children referred to popular songs which included the word 'Macedonia'. Similarly, children were unable to explain what a *'hero'* is and most of children referred to *'superheroes'* such as *'Superman'* and *'Batman'*. A small number of children were able to explain that heroes are the ones *'who win'* or *'who are brave'*.

Children's answers provided the necessary initial information for teachers and parents to locate the gaps in their understandings, or stereotypes (e.g., heroes are only men) and to start discussions around the Macedonian crisis/struggle topic. A deeper understanding of the challenge, and reframing and restructuring of initial plans took place at the end of this stage and then led to the second stage, namely 'interpretation'.

Interpretation

This stage was organised in two phases:

Phase One First, the topic was introduced. In this phase educational visits to museums, invited guests and educational materials from external sources offered some initial stimuli and information to enrich children's thinking. A field trip to the Museum of the Macedonian Struggle was organised where children had the opportunity to learn information and see exhibits related to that specific era. Each kindergarten received a 'training suitcase' from the museum with educational resources and relevant activity suggestions. These activities were implemented in the kindergartens (e.g., traditional puppet play with the heroes of that time;

exploring the swamp of Giannitsa as the central point of the struggle etc.). In addition, invited guests presented information and photographs of that period and explained how people lived, how they commuted, how they communicated, how schools operated and so on.

Phase Two After children had completed the activities mentioned above, they spent time discussing their new knowledge with their teachers and started brainstorming ways to find additional information. Children used books, comics, photographs, concept maps and narratives of elders who visited their schools. Parents were also involved in this stage. A letter was sent to parents to ask them for their cooperation in locating relevant information at home and to invite them to bring relevant information and artefacts to school.

During this stage it was evident that children's interest was extended, and their understandings significantly expanded. Stories and visits transformed children's views and insights and enriched their knowledge. Throughout this stage, extra attention was paid to avoiding stereotypes regarding the countries and also suggesting ways to resolve conflicts in a peaceful, just way. Rich conversations with teachers, parents and guests inspired children not only to learn more, but also to take action and find their 'clear direction for ideation' (IDEO Design Thinking for Educators, 2013, p. 39).

Ideation

Phase Three The third phase of the design model is all about brainstorming, generating and exchanging ideas. Children's views and ideas were encouraged as much as possible in this stage. They brainstormed information they would like to learn, questions they would like to ask and activities and actions they would like to implement. Collaborating and building on each other's ideas was also a necessary element of this stage. Visual clues (e.g., drawings and sketches) were also a vital part and helped with the processing and categorisation of the collected information and materials. They finalised their brainstorming with ways in which they could 'bring' the heroes of the Macedonian struggle into their own classrooms to help them find a peaceful solution to their 'struggle'. They decided to create a scenario for dramatic play and to construct puppet figures of the heroes who participated in the Macedonian struggle. Digital technology (eShadow software) would help them to convert their drawings into animated figures for their dramatic play on the screens of computers or during a live performance. A local cultural group presented traditional clothes and dresses people wore at that time, in order to enhance the accuracy of children's representations of people of that period. Children had the opportunity to discuss the different materials and also to wear some of those traditional clothes.

In one of the kindergartens, children created a timeline to follow the events and the facts of the Macedonian struggle. In small groups, they also drew paintings or created representations of the swamp of Giannitsa (the main site of the Macedonian struggle—thus incorporating the arts).

Children's new knowledge about people and places (integrating science and mathematics) augmented their understanding but also assisted with children's efforts to design their paper-based and digital puppets, bringing their creative ideas to life in the fourth stage: experimentation. Digital technologies and arts played a fundamental role in the next stage.

Experimentation

Phase Four Children's creative ideas were made tangible in this phase. Ideas were actualised and were shared with other people. Feedback from teachers, parents and the community helped with the refinement of prototypes (e.g., drawings, paper-based and digital puppets).

This fourth phase included the construction of the figures for the shadow puppet theatre and the writing of scenarios for their dramatic play. All the participating kindergarten classrooms were 'brought together' through Skype. Virtual conversations through Skype helped children to decide which figures and characters they would like to draw and then transform those into digital puppet figures. Shared leadership between children was evident in this phase as they had to make decisions about who was making what and for what reason. This was an important moment in the study as children had to justify their decisions to their peers and

Fig. 2.1 Paper-based figure of one of the most popular heroes: Pavlos Melas

also needed to persuade others about their ideas (e.g., which heroes they would choose, their role in the script etc.). Time constraints required children to make decisions quickly (e.g., who was drawing which hero). Via Skype, they also had the opportunity to discuss and finalise their scenario for the final performance.

Children first drew their protagonists, the main characters of their scenario both Greeks and Bulgarians, on paper and then constructed their paper-based puppet figures (Fig. 2.1) before digitising them with eShadow (Fig. 2.2).

Phase Five The fifth phase included activities relating to shadow theatre (the traditional form of shadow theatre) and the introduction of the eShadow software (http://www.eshadow.gr) to children. Shadow theatre is a storytelling tradition common to many countries in Asia and the Middle East using flat articulated puppets which are held between a light source and a translucent screen or a scrim. It is a medium with significant educational value (Hatzigianni et al., 2016) within the wider context of drama and performance arts. eShadow enriches the features of traditional shadow theatre with digital technology elements to offer a new way of dramatised and personalised digital storytelling. It enables the production of rich multimedia content, interactively using innovative input

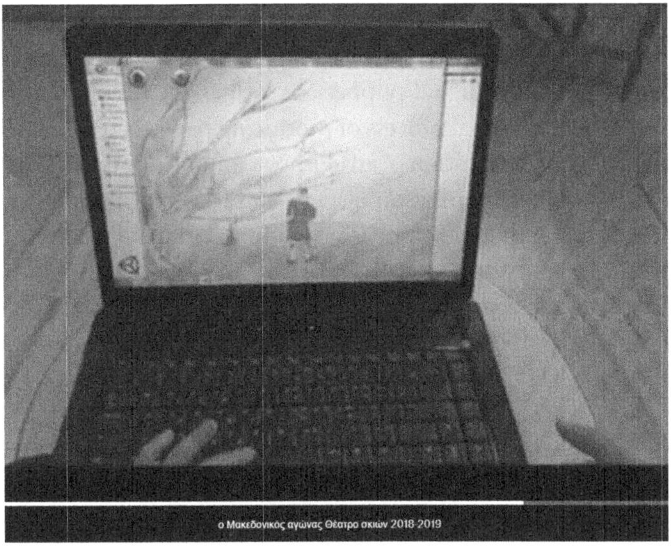

Fig. 2.2 The digital figure of Pavlos Melas

devices and it supports online collaboration. It offers an intuitive way of setting up scenes and enacting them. The user can select desired scenery objects and digital puppets and then move them with mouse drag operations. All movements can be easily recorded along with the voice of the user. These recordings can be exported in appropriate file formats to be further edited with external video processing tools.

eShadow emphasises the realistic motion simulation of shadow theatre puppets. Realistic movement engages children and promotes various learning activities (Moumoutzis, Christoulakis, Christodoulakis, & Paneva-Marinova, 2018). Furthermore, realistic movement of digital puppets creates an atmosphere of playful interaction where the users are very easily engaged in theatrical improvisations. This encourages the development of communication skills related to oral expression and interpretation of body language.

Although eShadow has been used for several years in primary and secondary schools for supporting student projects, its use in kindergarten presented some new challenges. Young children did not yet have the digital skills to create their own digital puppets using image processing software, as it was done with older children, usually with the help of a computer science teacher (Moumoutzis et al., 2017). The need to support a simpler and more intuitive way of developing new digital puppets, as well as the need to combine this process with the traditional process of producing paper-based puppets, led to the design and implementation of an ePuppet, an 'app' for mobile devices that is able to facilitate the digitisation of two- and four-part puppets, which are the most common types of puppets in traditional shadow theatre.

To introduce eShadow and ePuppet to the participating teachers, a workshop was held and several online meetings were conducted. After familiarising themselves with the software, the teachers introduced it to the participating children. Ongoing support was provided by the technical team to address any technical problems and to refine the ePuppet app in order to ensure its usability and effectiveness during the process.

Children were given ample opportunities to experiment with the software and the app. By the end of this phase, the digitisation of the puppets and the preparation of the digital play were both completed. In addition, children practised moving and 'acting' with their figures in the eShadow environment.

Dissemination activities were organised by children in order to present their work and to demonstrate their competent use of the software. The digital play was video recorded and presented to parents and the community as well as to children from other classrooms/kindergartens. Presentations took place at the end of the school year (June 2019, Fig. 2.3).

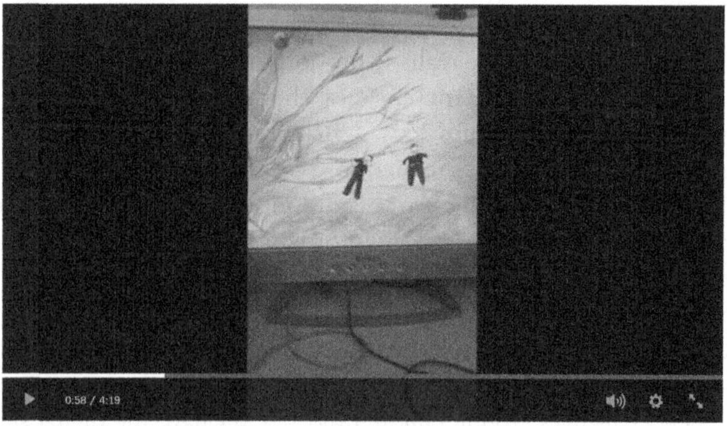

Fig. 2.3 Children play with the eShadow puppets on a computer

Evolution

Phase Six In this final stage, children were able to collect memories by reflecting on their favourite moments and any surprises or challenges they had experienced. Through this process, children were able to build a narrative; an overview of their thoughts, impressions and experiences. These narratives were also shared amongst the participating kindergartens and the broader community.

The last phase of this project included reflections and the gathering of feedback from internal and external sources. The same questions were again posed to all participating children during semi-structured interviews to identify children's *new* knowledge, impressions and understandings. Discussions regarding the value of peace and the knowledge children had gained about that historical period took place during this final stage. Children were now able to articulate what the Macedonian struggle was about. Their views about heroes were much more realistic and no reference was made to superheroes. For example, Sunny (four years old) was able to provide more depth and detail in her answer and explained that the Macedonian struggle was '*when Macedonians fought with Bulgarians who wanted to take Macedonia so that they could have an exit to the sea*'; Olga (four years old) was also very specific in her response: '*the Greeks from Macedonia fought with Bulgarians*'. Heroes, according to children, were now ordinary, real people who '*save others*', '*win*', '*fall into battle*', '*are brave like Pavlos Melas*' (Paul, 5 years old). Gendered perceptions of heroes as always male had evolved as well: children also referred to women's and children's roles in the struggle and how everyone helped each other, having a strong sense of belonging and community.

Children's views on learning to use eShadow, their experiences of performing a digital play, and the skills they had acquired and rehearsed were also explored. Children enjoyed using eShadow and were eager to use it again. With the support of their teachers, children reflected on the overall progress of the project and identified areas that they would do differently if they could do it again (e.g., having more time to work on the puppets, visiting the children at the other kindergartens and making the puppets together). The digital documentation of their presentations helped with their reflections at this stage.

As June 2019 marked the end of the school year in Greece, teachers could not plan further steps with the older children (the five-year-olds). These children would start primary school the following September. Teachers reported that when they returned to school in September 2019, after the school holidays, the younger children who had not transitioned to primary school were eager to work on eShadow software again and started planning next steps. The creation of mind maps about future ideas and questions children had for following projects contributed to starting the whole process again, proving sustained interest despite a three-month break. The project had thus contributed to building the foundations for a strong community of learners and future designers.

Implications and Conclusion

Through this design thinking process, young children's understanding of a real-world issue that is relevant to their everyday lives was deepened. The challenge of intentionally teaching complex, abstract concepts to young children was successfully addressed. More importantly, children had the chance to participate actively and creatively at every stage of the project. By implementing the design thinking approach, teachers did not separate knowledge in boxes of science, technology, engineering, the arts and mathematics. They worked holistically, taking advantage of every teachable moment, but also purposefully planning all learning experiences. Children's knowledge in all areas was advanced. Children explored their local environment (e.g., the swamp—science), learnt about clothes and materials that people used in different eras (materials—engineering) and the technology that was available 100 years ago (e.g., no cars, no Skype, no telephones—technology), experimented with different artistic techniques (e.g., natural materials; collage, water paint.—arts), created timelines and diagrams (mathematics) and also enhanced their digital skills (Skype, eShadow, ePuppet—digital technology).

The adoption of a design thinking approach in this study had numerous advantages. It facilitated a smooth integration of digital technologies and the arts but also helped children understand the continuity of learning. Children experienced organised brainstorming and careful design through which they were able to address challenges (e.g., understanding abstract concepts), solve problems (e.g., convert paper-based puppets into digital, animated puppets; sharing their stories and plays) and increase their impact on their community (e.g., by understanding the value of peace and justice). Activities stemming from a design approach, such as flow diagrams, process diagrams, paper prototypes, maps, two by two grids, storyboards, storylines, sketches and many others, could be particularly useful for young children. These activities made children's learning visible, underlining the value of process over product. Design thinking in early childhood also enabled collaboration, teamwork and the active involvement of parents, families and the local community.

Educators in this study were given broad directions around the IDEO model and the different stages but overall they were free to adjust their plans and activities according to children's needs and interests, and also according to their local context. The design thinking approach provided educators with enough flexibility to make them feel trusted, confident and relaxed throughout the process. They knew that certain activities should be enacted (e.g., the interviews and the eShadow puppets) but were at liberty to decide which artistic techniques they would use and whether or not to organise a large performance at the end of the school year. In addition, the fact that there was no testing or measuring of specific skills was also seen positively by the educators.

A recommendation for future projects would be to spend adequate time in locating a worthy challenge with the children: a challenge which would lead to the production of original, authentic and practical ideas and solutions. Plenty of time is required when a design approach is implemented, ideally more than one year, so that Phase 6: Evolution is sufficiently addressed. The educators' role is to guide and facilitate the whole process but also to provide feedback and cultivate a classroom culture in which children are encouraged to make, and learn from, mistakes and trials (Swanson & Collins, 2018). Metacognitive strategies and analogies between the problems they deal with and other problems that might be faced in other similar situations/contexts would enhance children's higher-order thinking and critical skills. For example, what would Bulgarian children think of Greek children's eShadow plays?

Overall, design thinking models have mostly concentrated on older children. More research and support are needed for successfully implementing them with young children. This study showed that early childhood education can easily accommodate design thinking approaches due to its flexibility, multimodality and use of open-ended environments.

Recommendations for Educators and Parents

Explore and Identify Alternative Design Thinking Models for Your Problem There are different design thinking models for education (see Appendix, p. 38). This is an emerging field which keeps growing. Reviewing the models available would give teachers/parents a more accurate idea of which to utilise for their project and for their specific context. Models with a strong educational background and with a range of free resources would be more suitable and easier to implement.

Build Home–School–Community Partnerships Design thinking is particularly useful for teamwork, collaboration and for projects which require parents and educators to work together to solve a local (school or community) problem/issue. Parents and community experts can have active roles within the broader project, based on their diverse expertise. Educators can coordinate the project, drawing on their leadership and pedagogical skills. Adults modelling high levels of work ethic and building collaborative teams would be an invaluable lesson for children.

Careful and Meticulous Co-ordination and Organisation Design thinking is a holistic approach, valuing authentic learning and process over product. Children's engagement and interest are almost guaranteed. However, the role of the adults in encouraging children, coordinating their work and encouraging sustained attention is crucial. Failures and mistakes are a necessary part of solving problems and achieving success. Children are encouraged to keep trying and practise resilience strategies. Instilling confidence, positive dispositions and aspirations around

STE(A)M disciplines is as vital as enhancing knowledge. Working intentionally to eliminate any gender stereotypes (e.g., boys are not good at paying attention; girls are not good at technology) and to promote digital equity (e.g., being mindful that some children do not have access to digital devices at home) is also critical in the early years.

Provide Plenty of Time for the Whole Project and Particularly for Evolution and Reflection (Metacognitive Skills) The last phase of any design thinking model is reflection and being the last phase of a project, is sometimes rushed. To avoid this rush, plenty of time should be allocated to this phase right from the start. Children should have the opportunity to think back and evaluate the initial phases of their project, but they also have the chance to think ahead and plan ways in which to improve effectiveness and efficiency.

Strengthen Digital Literacy Skills The integration of digital technology in this project was very successful. Depending on the topic of the project, the age of the children, and the school/home resources, a variety of digital tools could be incorporated. Some examples include survey software (e.g., SurveyMonkey); brainstorming and concept maps (e.g., Kidspiration Maps; Inspiration; CmapTools; Graphmind; Text 2 Mind Map; Coggle); 3D printers; 3D design apps (e.g., Makers Empire; Tinkercad; SketchUp); whiteboard apps and software (Stormboard; Mural); cloud technologies (e.g., Google suite; Dropbox); storyboard and presentation apps (e.g., Prezi Next, Canva, Seesaw, Popplet); Robotics; Programming (e.g., Scratch, ScratchJr); Internet of Toys (of Things); digital laser cutters and laser engraving machines; AR and VR headsets and apps and many others.

Acknowledgements We would like to acknowledge the hard work and effort made by all the teachers, children and parents in order for this project to be completed. Without their contribution, this study would not have been possible. Our special thanks to the teachers of the kindergarten who were directly involved: Ioanna Karagiorgou, Despoina Kaladaridou, Alexia Panta, Euthemia Gioti, Aspasia Zeibeki and Vicky Alexiou.

Appendix

Design thinking coding—Actions and example dialogue for all children. Available from: https://primary-makers.com

Model/stage	Early		Mid		Late		
Cooper Hewitt (Hewitt, 2011)	Identify	Investigate	Frame/ reframe	Generate	Develop	Evaluate	Re-evaluate
d.School (Hasso Plattner Institute of Design, 2017)	Empathise	Define	Ideate		Prototype	Test	
Design Minds (State library of Queensland, 2017)	Inquire	Reflect	Ideate	Reflect	Implement	Reflect	
IDEO (Fierst et al., 2011)	Discovery	Interpretation	Ideation	Experimentation	Evolution		
IDESiGN (Burnette, 2005)	Intending	Defining	Exploring	Suggesting	Innovating	Goal-setting	Knowing
NoTosh (Mcintosh, 2018)	Immersion	Synthesis	Ideation		Prototyping	Feedback	
Open Colleges (Briggs, 2013)	Define the problem		Consider multiple options	Refine selected direction	Execute the best plan of action		

References

Bevan, B. (2017). The promise and the promises of making in science education. *Studies in Science Education, 53*(1), 75–103.

Bower, M., Stevenson, M., Falloon, G., Forbes, A., & Hatzigianni, M. (2018). *Makerspaces in primary school settings – Advancing 21st century and STEM capabilities using 3D design and 3D printing*. Sydney, Australia: Macquarie University. Available at https://primarymakers.com

Briggs, S. (2013, July 29). 45 Design Thinking Resources for Educators. Retrieved April 27, 2021, from https://www.opencolleges.edu.au/informed/features/45-design-thinking-resources-for-educators/

Buchanan, R. (1992). Wicked problems in design thinking. *Design Issues, 8*(2), 5–21.

Burnette, C. (2005). IDESiGN. Retrieved April 27, 2021, from http://www.idesignthinking.com/main.html

Crockett, L. (2011). *Literacy is not enough: 21st-century fluencies for the digital age*. Thousand Oaks, CA: Corwin.

Exarchate, B., Mazarakis-Ainian, K., Koromilas, L., Demestichas, I., Katechakis, G., Dragoumis, I., & Petkov, A. *Macedonian struggle*. Retrieved December 9, 2019., from https://infogalactic.com/info/Macedonian_Struggle

Fierst, K., Diefenthaler, A., & Diefenthaler, G. (2011). *Design thinking for educators*. Riverdale, CA: IDEO.

Filatro, A., Cavalcanti, C. C., & Muckenberger, E. (2017). *Design thinking and online education*. Retrieved September 20, 2019, from https://digitalcommons.andrews.edu/adventist-learn-online/2017/tuesday/12/

Hasso Plattner Institute of Design. (2017). A Virtual Crash Course in Design Thinking. Retrieved April 27, 2021, from https://dschool.stanford.edu/resources-collections/a-virtual-crash-course-in-design-thinking

Hatzigianni, M., Miller, M., & Quiñones, G. (2016). Karagiozis in Australia: Exploring principles of social justice in the arts for young children. *International Journal of Education & the Arts, 17*(25), 1–20.

Hatzigianni, M., Stevenson, M., Falloon, G., Bower, M., & Forbes, A. (2020). Children's views on making and designing. *European Early Childhood Education Research Journal, 28*(2), 286–300. https://doi.org/10.1080/1350293X.2020.1735747

Hewitt, C. (2011, September 9). Ready, Set, Design! Retrieved April 27, 2021, from https://www.cooperhewitt.org/2011/09/09/ready-set-design/

Honey, M., Pearson, G., & Schweingruber, H. (2014). *STEM integration in K-12 education: Status, prospects, and an agenda for research*. Washington, DC: National Academies Press.

IDEO Design thinking for educators. (2013). Retrieved March 21, 2018, from https://www.ideo.com/post/design-thinking-for-educators

Kermani, H., & Aldemir, J. (2015). Preparing children for success: Integrating science, math, and technology in early childhood classroom. *Early Child Development and Care, 185*(9), 1504–1527.

Koukoudakis, G. (2018). The Macedonian question: An identity-based conflict. *Mediterranean Quarterly, 29*(4), 3–18.

McIntosh, E. (2018, April 17). Think differently and change the way you choose to work [text/html]. Retrieved May 29, 2018, from https://notosh.com/search/results?q=McIntosh

Moomaw, S., & Davis, J. A. (2010). STEM comes to preschool. *YC Young Children, 65*(5), 12–18.

Moumoutzis, N., Christoulakis, M., Christodoulakis, S., & Paneva-Marinova, D. (2018). Renovating the cultural heritage of traditional shadow theatre with eShadow. Design, implementation, evaluation and use in formal and informal learning. *Digital Presentation and Preservation of Cultural and Scientific Heritage Proceedings, 8*, 51–70.

Moumoutzis, N., Christoulakis, M., Pitsiladis, A., Maragoudakis, I., Christodoulakis, S., Menioudakis, M., & Tzoganidis, M. et al. (2017, April). Using new media arts to enable project-based learning in technological education. In 2017 *IEEE Global Engineering Education Conference (EDUCON)* (pp. 287–296). Piraeus, Greece: IEEE.

Shively, K., & Stith, K. (2018). Measuring what matters: Assessing creativity, critical thinking, and the Design Process. *Gifted Child Today, 41*(3), 149–158. https://doi.org/10.1177/1076217518768361

Sofou, E., & Tsafos, V. (2010). Preschool teachers' understandings of the national preschool curriculum in Greece. *Early Childhood Education Journal, 37*(5), 411–420.

State Library of Queensland. (2017). Design Minds | Design Online. Retrieved April 27, 2021, from http://designonline.org.au/education/

Stevenson, M., Bower, M., Falloon, G., Forbes, A., & Hatzigianni, M. (2019). By design: Professional learning ecologies to develop primary school teachers' makerspaces pedagogical capabilities. *British Journal of Educational Technology, 50*(3), 1260–1274.

Swanson, H., & Collins, A. (2018). How failure is productive in the creative process: Refining student explanations through theory-building discussion. *Thinking Skills and Creativity, 30*, 54–63. https://doi.org/10.1016/j.tsc.2018.03.005

Zahariadis, N. (1994). Nationalism and small-state foreign policy: The Greek response to the Macedonian issue. *Political Science Quarterly, 109*(4), 647–667.

3

Creative Digital Art: Young Children's Video Making Through Practice-Based Learning

Suzannie K. Y. Leung, Kimburley W. Y. Choi, and Mantak Yuen

Introduction

In this chapter, we focus on digital art as encompassing both the A (the arts) and the T (technology) in STEAM education. In the twenty-first century, digital play is part of young children's lives. Most young children grow up in technology-rich homes, and they engage in frequent digital

S. K. Y. Leung (✉)
Department of Curriculum and Instruction, The Chinese University of Hong Kong, Hong Kong SAR, China
e-mail: suzleung@cuhk.edu.hk

K. W. Y. Choi
School of Creative Media, City University of Hong Kong, Hong Kong SAR, China
e-mail: smkim@cityu.edu.hk

M. Yuen
Faculty of Education, The University of Hong Kong, Hong Kong SAR, China
e-mail: mtyuen@hku.hk

C. Cohrssen, S. Garvis (eds.), *Embedding STEAM in Early Childhood Education and Care*, https://doi.org/10.1007/978-3-030-65624-9_3

play. Video art has also become increasingly popular in children's worlds. Children can capture moving images creatively using a video recorder. This chapter uses a video-making project with young children as a case study to discuss the potential value of creative digital art in early childhood education. The implementation processes are described for this successful workshop, along with other possible activities for toddlers and children, to undertake with teachers and parents.

Digital Play in Early Childhood Arts

'Art' in early childhood refers to original expression through dance, drama, visual arts and music (Bresler, 1998). The arts are crucial for children's creative development in a play-oriented early learning context (Wright, 2007). Creative behaviours and visual patterns are usually found in early childhood play activities (Arnold, 2010), and the symbiosis between creativity, art and play provides precious learning opportunities for children in their early years (Wright, 2014). Art-making creates learning opportunities for children to elaborate on their thoughts in a non-verbal way and serves as a platform for meaningful play (Korn-Bursztyn, 2002). It is widely believed that children should be exposed to every art form in a balanced fashion, since each form makes a specific positive contribution to children's development and learning (Bautista, Moreno-Núñez, Koh, Amsah, & Bull, 2018; Gadsden, 2008; Hanna, 2014).

Along with fine-art genres such as drawing, painting and sculpture, digital art is a new genre of contemporary arts and belongs to the spectrum of visual arts in early childhood education. Digital arts use digital technology as part of the creative process. However, the place of digital technology in early childhood has been controversial in the last two decades. For example, some researchers have suggested that using digital technology at a young age may cause children to experience negative outcomes in three main areas: health and well-being, cognition and brain development, and social and cultural competencies (Bolstad, 2004; Stephen & Plowman, 2014). Nevertheless, it seems that digital play is almost an unavoidable experience for young children in this age of technology. Studies in several countries have illustrated that children are growing up in technology-rich environments, with digital devices such as tablet computers, smartphones, cameras and video recorders already part of typical life at home (Fleer, 2013; Livingstone & Haddon, 2014; Plowman, 2015). In light of this, some educators and researchers have argued that digital technology should be utilised to support children's learning in different areas such as literacy (Marsh, 2012; Nash, 2012), the arts (Terreni, 2011) and mathematics (Jowett, Moore, & Anderson, 2012).

Vygotsky (1997) theorised that children's play and learning involve the sociocultural concept of tool mediation. Tools in this sense are derived from the cultural contexts in which people participate, and tool mediation refers to people's use of tools to achieve a particular objective during an activity (Vygotsky, 1997). This description seems to apply particularly well to the relationship between children's play and the use of digital devices. Children's higher-order mental functions that involve conceptual development and logical thinking, and lower-order mental functions such as memory and attention, are essentially supported and mediated by the use of cultural tools (Daniels, 2008).

Hutt (1979) has suggested that children may change and adjust their play behaviours according to their current experiences. Hutt's subsequent work with her colleagues involved studying young children's play with novel objects (Hutt, Tyler, Hutt, & Christopherson, 1989). They developed a taxonomy of play, based on two categories: epistemic play and ludic play. *Epistemic* play occurs when children explore objects and environments to acquire knowledge. *Ludic* play happens when children manipulate or employ an object to create symbolic, innovative or imaginative play according to their learned experiences. More recently, Bird, Colliver, and Edwards (2014) drew on Hutt's (1979) taxonomy of children's play with novel objects and Vygotsky's (1997) theory of the cultural basis of knowledge formation to develop a "digital play framework". By showing how children systematically learned to use a still camera through play, Bird et al. (2014) describe a set of behaviours that align with epistemic and ludic activity. They showed that children demonstrating epistemic play tend to engage in exploration, problem-solving and skill acquisition, while those involved in ludic activities are likely to participate in symbolic or innovative play. Subsequently, Bird and Edwards (2015) extended this work to develop a broader digital play framework, in an attempt to understand how children learn to use various kinds of technologies through play.

The Arts in Early Childhood: The Hong Kong Context

Over the past decade, early childhood education in Hong Kong has undergone reform and there has been a dramatic change in curriculum. As a result of the reform and the required changes in approach that it brings, many kindergartens have faced considerable challenges because curriculum changes have clashed with the traditional expectations of parents. Traditionally, the programme in early childhood settings had been influenced by an achievement-driven Chinese culture, with an emphasis on formal instruction for young children (Cheng, 2001; Rao, Ng, & Pearson, 2010). Time and resources were devoted mainly to academic activities to ensure children's school readiness. Prior to these curriculum changes, early childhood educators in Hong Kong placed emphasis mainly on language and cognitive development (Li, Rao, & Tse, 2012; Ng & Rao, 2008) and very limited opportunities were provided for kindergarten children to engage in creative arts experiences (Bautista et al., 2018; Leung, 2018). When offered, visual arts activities were mainly limited to craft making since arts-related pedagogical practices in Asian kindergartens tend to be product-oriented by nature (Bautista et al., 2018).

The Education Bureau in Hong Kong advocated for a much stronger emphasis on respecting and fostering children's uniqueness and creativity. The *Kindergarten Education Curriculum Guide* was launched in 2017, positioning "arts and creativity" as one of the core learning areas in the preschool curriculum (Curriculum Development Council, 2017). Today, the importance of arts education within balanced kindergarten, primary and secondary curricula in Hong Kong is recognised. Further, the arts have been combined with Science, Technology, Engineering, and Mathematics (STEM) to become STEAM (Maeda, 2013; Reitenbach, 2015; Sousa & Pilecki, 2012).

Innovations in arts education continue in Hong Kong. However, whilst numerous overseas studies have shown that arts experiences have a strong positive influence on children's learning (Menzer, 2015), little research has focused on early childhood art in the context of early education in Hong Kong. The role of digital devices in educational contexts has similarly not yet been fully investigated (Fleer, 2016; Marsh, Plowman, Yamada-Rice, Bishop, & Scott, 2016). Drawing on both arts education and digital technology, the first author set out to investigate children's learning about photographic art and recommended that kindergarten teachers need to be equipped with digital arts knowledge and skills to enhance children's creativity (Leung, 2014). Kindergarten teachers also reported that visual arts teaching is difficult because they had limited evidence-based training in visual arts education (Leung, 2017).

Moving Images for Young Visual Thinkers: A Case Study

The Centre for Advancement in Inclusive and Special Education (CAISE) was established in February 2004 in The University of Hong Kong's Faculty of Education. The primary aims of the centre are to advance research and services in special education, and to encourage cross-disciplinary research collaboration in the field of special education.

A two-day video-making workshop was designed and included in a summer holiday creativity and talent development programme. The aims of this workshop were to equip children to (1) demonstrate shooting and editing skills to convey a message through the medium of video, (2) use symbolic moving images to articulate their thoughts and viewpoints and (3) use film language in their videos to demonstrate their aesthetic and creative appreciation. The two-day workshop taught the children cinematic language through video making, technical skills through operating a video camera and narrative techniques through creating storyboards. With the support of the School of Creative Media at City University of Hong Kong, workshop resources included technical support, seven professional video cameras, four hand-held video cameras, three tripods for recording moving images and computer notebooks for editing.

Nine primary school children (aged five to eight years; one boy and eight girls) participated in a workshop entitled Moving Images for Young Visual Thinkers. The children elected to participate in the workshop in which each child would create a short movie, selecting the workshop from a list of options available to them during the summer break. The children were from local families and as they all spoke Cantonese and English, the workshop was conducted bilingually.

Two of the authors of this chapter led the workshop with the assistance of four facilitators. Both instructors are knowledgeable in film language and techniques and one is an early childhood education specialist. We were interested in how the children would interact with digital devices to achieve a particular outcome, and how educators should support digital learning for children. The four facilitators each hold bachelor degrees in early childhood education and they were able to communicate easily with the children at an appropriate language level. Three alumni of the School of Creative Media served as movie editors for the children's final video works.

Since video art is a novel area in early childhood education, all the learning and teaching behaviours in this workshop were recorded and analysed as a research study. Parents were aware of the purpose of the research and signed a consent form for their child to participate in the study and to be recorded on video (without showing their faces) during the workshop. The video recording of the children interacting with the instructors and facilitators allowed the authors to observe and analyse the processes and actions that took place.

Digital Play Through Practice-Based Exploration and Guidance

The concept of moving images was explained on the first day of the video-making workshop by introducing a flipbook to the whole class as a 20-minute warm-up activity (Fig. 3.1).

Fig. 3.1 Flip book compilation by professional animators. (Source: https://www.youtube.com/watch?v=JVzf9rtgf9Y)

Teacher: What can you see in this video?
Participant: There is a book!
Teacher: What did you see in this book?
Participant: The drawings! They are moving in the book.
Teacher: People try to use a flip book to show moving images. But nowadays, since we have digital cameras, we can create moving images through the machine.

The teaching team had created a tailor-made, child-appropriate 42-second video to introduce concepts such as filler effect, long take, on- and off-screen, point-of-view shot and zoom-in/zoom-out effects. One of the facilitators, Winnie, and her friend, Winnie the Pooh (Fig. 3.2), took the lead roles in this video, which helped to engage the children with the content of the video. After the children had watched this video twice, they were asked some questions about the basic features of a video.

Fig. 3.2 Training video created by the teaching team to introduce the daily-life experiences of Winnie and Winnie the Pooh

Teacher:	What can you see in this video?
Participant:	Winnie and Winnie the Pooh!
Teacher:	Are they standing still or moving?
Participant:	They are moving.
Teacher:	What kinds of colour do you see in this video?
Participant:	Orange, black and grey.
Teacher:	Did you hear any sound from this video?
Participant:	Winnie is talking to the Pooh.
Teacher:	Yes, now we know that a video has a cast, objects, sounds and even a script so that people will know what it is about.

PowerPoint slides were then shared. These highlighted the concepts that had been introduced in the video (filter effect, long take, on- and off-screen, point-of-view shot and zoom-in/zoom-out effects). Specific shots were played back and paused so that they could be discussed. For

Fig. 3.3 Example of a PowerPoint slide created by the teaching team to illustrate filming techniques

instance, the children discussed the zoom-in/zoom-out effects with a slide and the video together (Fig. 3.3).

Teacher:	Could anyone tell us what a zoom-in effect is?
Participant:	A zoom-in is something that gets an object more closely, and that thing will become very big.
Teacher:	(The teacher came closer to the children and shortened the distance.) Is this a kind of zoom-in?
Participant:	Yes!
Teacher:	(The teacher stepped back from the children.) How about now?
Participant:	This is zoom-out!

This training session was followed by an exploratory session to allow the participants to try out the video devices and techniques using a professional video camera. None of the children had seen a professional video camera before. The children formed groups of three, with one facilitator supporting them conceptually and technically. Each group was asked to draw two cards, each reflecting a videoing technique they had learnt about in the previous session. With the support of a facilitator, the children then made short videos that showed these two techniques (three minutes for each technique). This exploratory activity lasted around 30 minutes. The finished videos were connected and projected on the screen one by one.

Teacher:	How nice! What kind of effect is this? Who can make a guess?
Participant:	I know! He talks to the sky, so this is a point-of-view shot.
Teacher:	Exactly! Let's clap our hands for the video maker! Who did this nice shot?

Later on, in another session, the children had 60 minutes to create a video scene as a whole-group activity using the professional cameras independently. Props provided by the teaching team included chopsticks, spoons, stainless steel plates, fruit (bananas and oranges), towels, umbrellas, paper boxes, glasses, caps and hats. In one scene, the children used a range of techniques to record symbolic play in which they were crossing a river in a boat. The children who did not have an acting role served as crew members to record the scenes and to show video recordings to the children who had acted in the video.

Creative Narration in Storyboard Drawing

In order to draw on children's lived experience, emails sent one day before the workshop invited each child to bring one or two of their most precious possessions or toys from their home with them to the workshop. This activity was called My Best Friend. The children used their toys to brainstorm their video ideas and start the storyboard drawing process. They asked questions such as the following:

"What is the toy's name?"
"What does he look like?"
"What kind of food does he like to eat?"
"What does he like to do?"
"What do you like to do with him?"

Fig. 3.4 Emilie's storyboard describing Elsa's singing and dancing

The children were invited to share their stories with the other children. After the sharing session, the children were divided into three groups. Each group was supported by one facilitator, who helped the participants with the video techniques as well as with the language to explain the events and describe the characters (Fig. 3.4).

Video Art in a One-Minute Movie

During the second day of the workshop, the children used their storyboards to shoot their videos, supported by adults who held the cameras and handled the props. The children played the role of film directors, while the adults acted as production assistants. The children shot all the footage they wished to have according to their storyboards. The facilitators then helped them record voiceover clips to add to the visual footage. Two professional editors helped the children edit their footage into a one-minute video. The participants' parents were invited to attend a screening later that afternoon, in which the "directors' dialogues" were shared with the audience.

Operating the video camera and performing most of the film techniques was challenging to Emilie (aged five) as she had small hands. However, her fingers could reach the zoom buttons and initially she zoomed in and out without apparent purpose. Emilie chose to focus on her doll, Elsa (the main character of a famous movie), for her storyboard. Emilie drew a picture to be a stage for Elsa, with snow and a castle. She learned to place the camera by moving a tripod to an appropriate position and, demonstrating epistemic play, located the correct angle so that the stage would fit into the camera shot. She demonstrated the techniques of zoom-in and zoom-out to ensure her desired composition using the stage she had created as the background for a theatre in which Elsa could sing and dance. While she was singing and manipulating Elsa's body, the facilitators held and operated the camera for her (Fig. 3.5). Emilie put on headphones, sang the song in the cartoon and edited this into her video as an audio clip:

Let it go. Let it go.
I'm one with the wind and sky.
Let it go. Let it go.

Fig. 3.5 Emilie sang the song "Let It Go!" as she manipulated Elsa's body so that she could dance

The perfect girl is gone.
Here I stand and . . . I don't care.
Let the storm rage on.
The cold never, never bothers me anyway.

Implications

Digital Play in Early Childhood

In this study, we observed how the children learned to explore and use a novel instrument (a professional video camera) and supportive equipment (the tripod and headphones) intentionally. Although it is almost impossible to offer an absolute definition of play (Hännikainen, Singer, & van Oers, 2013), this study showed how the children performed recognised forms of play associated with children's early years. They explored, manipulated, pretended and even innovated, using the digital devices to create video works. The children's epistemic play experiences extended to ludic play activities using props and cinematographic techniques to create a one-minute video.

By recording the children's behaviour across the duration of the workshop, we are able to contribute further indicators associated with learning to use video cameras through play (Leung, Choi, & Yuen, 2020) to Bird et al.'s (2014) digital play framework. This case study also offers teachers new insights into how children learn to use technologies through play. In particular, the new indicators provide guidance on how to observe, plan and integrate the use of digital technologies in play-based learning in early childhood education (Table 3.1).

Facilitators' and Instructors' Perspectives: What Did We Need to Know?

What do teachers need to know in order to teach? Teaching a subject does not merely require teachers to acquire the necessary content knowledge; they also need to understand *how* to deliver the subject matter in such a

Table 3.1 New indicators relating to digital play based on the digital play framework

Types of digital play	Specific behaviours	Existing indicators in the digital play framework (Bird et al., 2014)	New indicators (Leung et al., 2020)
Epistemic play	Exploration	Holding the camera upright	Moving the tripod around
		Locating the start/stop button	
		Orientating the viewfinder	
		Seemingly random footage	
	Problem-solving	Pressing the start/stop button	Using a phone's ringtone as background music
		Turning the camera in relation to what is seen in the viewfinder	Placing the colour filter on top of the camera
		Intentional but "uncontrolled" footage	
	Skill acquisition	Framing footage in the viewfinder	Tilting the tripod up and down
		Using the zoom function	Using headphones to listen to the sound recording
		Intentional and controlled footage of observable people, events and situations	
Ludic play	Symbolic	Deliberate footage of peers involved in play	Deliberate footage of improvising with props in pretend play
		Deliberate footage of pretend play established for the purpose of filming	
	Innovation	Deliberate filming of content generated for the purpose of creating footage	Deliberate filming of content by using film language in the narration

way that the learner is supported to learn (Shulman, 1986). This requires the teacher to know what came before the new knowledge and what will be built on it, how it sits within a wider learning context, what may be difficult for the child to understand and how to adapt the teaching process to avoid such difficulties. In other words, teachers must also acquire "pedagogical content knowledge" (Shulman, 1986, p. 9).

After the workshop, the researchers conducted a reflection meeting with the facilitators and instructors. The facilitators shared how they perceived their role in the children's creative digital play in this workshop. Their reflections revealed the importance of content knowledge (digital media) and pedagogical content knowledge (how to support children learning about digital media in developmentally appropriate ways) for teachers involved in digital play. Facilitators and instructors with different educational backgrounds coached children differently from facilitators and instructors with digital media backgrounds. Those with early childhood education backgrounds had less content knowledge and experienced difficulty in guiding the children to use storyboards to tell their stories. Their lack of film arts education also hindered their attempts to facilitate the children's making of digital moving images, although they displayed a great eagerness to teach. Clearly, if digital media, as one form of technology education, are to be included in early childhood curricula, then digital media education must be incorporated in pre-service teacher education programmes (Table 3.2).

Affordances of Digital Devices

Investigations into how kindergarten teachers perceive the affordances of different digital devices in visual arts education for young children are rare. This study sought to reveal how the affordances of a video camera may be affected by the device itself, the nature of the digital play and the competencies of the facilitators in guiding the children to create their video artworks. In a world in which children are spending increasing amounts of time using digital media for different purposes, the contribution of digital media and how it can be used to support child learning is an emerging area of early childhood research (Madanipour & Cohrssen, 2019) and one that is further addressed in two chapters of this book.

Table 3.2 Facilitators' and instructors' reflections on their teaching competencies in different kinds of digital play activities in the workshop

Activities in the workshop	Facilitator without training in film studies	Facilitator with informal training in film studies	Instructor with formal training in film studies
Digital explorative play	I was just able to instruct the children in simple functions, like how to press the record button and how to zoom in and out. I was not very confident teaching them to operate the video cameras.	I showed them how to use different buttons from the video camera to perform film language (e.g., point-of-view shot and long take). I was uncertain of the exact way to teach. I found difficulties in explaining the concepts.	I introduced the buttons with different functions to the children and then shot some trial shots and explained the effects to them through the viewfinder. I did my teaching naturally, and I had not planned the details in the teaching process.

(*continued*)

Table 3.2 (continued)

Activities in the workshop	Facilitator without training in film studies	Facilitator with informal training in film studies	Instructor with formal training in film studies
Storyboard drawing	I tried to motivate the children by saying that drawing a storyboard is a good way to express what they want to shoot for their videos clearly. However, it was very challenging for me to facilitate them in expressing the key idea in every single shot. I spent a lot of time thinking of how to explain to the children which part was important and which part was less important in order to finish a one-minute video.	It was not easy to break down their stories into different scenes and add effects to the scenes. This is not a common practice for me. It takes some time to imagine how the shots are going to be shot, to consider the selection of filming techniques that fit the scenario and the adoption of extra props, etc. I tried to show them what their toys were going to be like on screen with different effects.	We designed this course, from drawing a storyboard to shooting a video, as a digital learning process. We do the same in our basic training in film studies, and we transformed this as a kid's version in this workshop. This is a usual practice for most filmmakers.

(continued)

Table 3.2 (continued)

Activities in the workshop	Facilitator without training in film studies	Facilitator with informal training in film studies	Instructor with formal training in film studies
Video shooting	I found that the children were absolutely out of control. I was not sure what exactly I had to tell them to do, especially for those who lost the ideas in their mind. I realised that there was a big gap between their ideas and my facilitation. I had no idea what else I should suggest for them to do to create that video.	It was a difficult task, really. The ideas from their storyboards were rough. When I asked them about the details, they changed their mind, and it was different from the storyboard. Sometimes I could hardly catch up with their thoughts. Therefore, I had a quick discussion with them on how to shoot the scene and explained the feasibility of their ideas. It was very time-consuming, and we spent too much time on the discussions.	The children showed us their storyboards, but some of them could not tell us how exactly he/she wanted the shot to look like. I would try a shot and confirm with him/her by viewing the shots through the viewfinder.

Practical Recommendations

Here we offer practical recommendations for teachers and parents to consider when using digital media in play with very young children.

Clay Animation This activity, which integrates the elements of craft and animation, is suitable for both toddlers and children. Teachers or parents can invite children to create an object using clay. Let's use a snowman as an example. First, the lower body of the snowman is made. Then, the child takes a photo of the snowman. (Make a mark to indicate where the snowman is standing when the photo is taken.) The children then continue to create the middle part of the body and take another photo of the snowman at the same spot. After this, the children finish the upper part of the body and the other elements of the snowman (e.g., the buttons on the body), taking photos of the snowman as it develops. The children then insert the photos into a moviemaker app and connect the photos sequentially as an animation showing the growth of the snowman. For toddlers, teachers and parents can help put the photos into the moviemaker.

Video Theatre Toddlers and children could explore this activity in different ways. Place the camera where it cannot be knocked over. Connect it to a projector and leave it running. Toddlers develop self-concept as they observe themselves on the projection screen. They learn how to control their emergence on the screen. Children can bring toys or found objects from home and play with them creatively in front of the video camera, at the same time, providing a voiceover as storytellers. Showing these videos to other children, teachers or parents supports emerging child identity.

Close-up Shooting Teachers and parents could try this activity with toddlers and children to enjoy scientific exploration. By using a video camera to capture visual images, children are able to observe some scientific phenomena in their daily life experiences. Support children to set a video camera on a table and shoot a cup of water (a close-up shot). They could record the moment when sugar melts in hot water or two colours mix together. Watch the video together asking what, why, when and how questions to encourage children to engage in higher-order thinking about what may have caused this scientific process to occur. The children could also play with the fast-forward button to demonstrate the concept of time. This process lends itself to the use of a smartphone or a digital camera to capture any form of change such as a bud in a vase opening into a flower, or soap flakes and water becoming slime when mixed. The teachers and parents could apply time-lapse techniques to the changes with children.

Conclusion

Incorporating technology and the arts, digital art is an important genre in the contemporary arts and belongs to the spectrum of visual arts in early childhood education, along with fine art genres such as drawing, painting and sculpture. Although the concept of STEAM has been recognised in early childhood education, few studies have focused on early childhood arts in Hong Kong. Fewer yet have focused on creative *digital* art with young children. This chapter seeks to bring a fresh perspective, with the goal of expanding the arts education knowledge and pedagogies of kindergarten teachers and parents from fine arts to media arts.

References

Arnold, C. (2010). *Understanding schemas and emotions in early childhood.* London: Sage.

Bautista, A., Moreno-Núñez, A., Koh, S. F., Amsah, F., & Bull, R. (2018). Arts-related practices in preschool education: An Asian perspective. *Early Childhood Research Quarterly, 45*, 277–288.

Bird, J., Colliver, Y., & Edwards, S. (2014). The camera is not a methodology: Towards a framework for understanding young children's use of video cameras. *Early Child Development and Care, 184*, 1741–1756.

Bird, J., & Edwards, S. (2015). Children learning to use technologies through play: A digital play framework. *British Journal of Educational Technology, 46*(6), 1149–1160.

Bolstad, R. (2004). The role and potential of ICT in early childhood education: A review of New Zealand and international literature. Retrieved from http://www.nzcer.org.nz/system/files/ictinecefinal.pdf

Bresler, L. (1998). "Child art," "fine art," and "art for children": The shaping of school practice and implications for change. *Arts Education Policy Review, 100*(1), 3–10.

Cheng, D. P. W. (2001). Difficulties of Hong Kong teachers' understanding and implementation of "play" in the curriculum. *Teaching and Teacher Education, 17*, 857–869.

Curriculum Development Council. (2017). *Kindergarten education curriculum guide*. Hong Kong, China: Curriculum Development Council.

Daniels, D. (2008). *Vygotsky and research*. New York: Routledge.

Fleer, M. (2013). Digital positioning for inclusive practice in early childhood: The cultural practices surrounding digital tablets in family homes. *Computers in New Zealand Schools: Learning, Teaching, Technology, 25*(1–3), 56–76.

Fleer, M. (2016). Theorising digital play: A cultural-historical conceptualisation of children's engagement in imaginary digital situations. *International Research in Early Childhood Education, 7*(2), 75–90.

Gadsden, V. L. (2008). The arts and education: Knowledge generation, pedagogy, and the discourse of learning. *Review of Research in Education, 32*(1), 29–61. https://doi.org/10.3102/0091732x07309691

Hanna, W. (2014). A Reggio-inspired music atelier: Opening the door between visual arts and music. *Early Childhood Education Journal, 42*(4), 287–294. https://doi.org/10.1007/s10643-013-0610-9

Hännikainen, M., Singer, E., & van Oers, B. (2013). Promoting play for a better future. *European Early Childhood Education Research Journal, 21*(2), 172–184.

Hutt, C. (1979). Exploration and play. In B. Sutton-Smith (Ed.), *Play and learning: The Johnson and Johnson pediatric round table* (pp. 175–194). New York: Gardner Press.

Hutt, J., Tyler, S., Hutt, C., & Christopherson, H. (1989). *Play, exploration, and learning: A natural history of the preschool*. London: Routledge.

Jowett, E. L., Moore, D. W., & Anderson, A. (2012). Using an iPad-based video modelling package to teach numeracy skills to a child with an autism spectrum disorder. *Developmental Neurorehabilitation, 15*(4), 304–312.

Korn-Bursztyn, C. (2002). Scenes from a studio: Working with the arts in an early childhood classroom. *Early Childhood Education Journal, 30*(1), 39–46.

Leung, S. K. Y. (2014). How does photography express children's voice? *Every Child, 20*(4), 16–17.

Leung, S. K. Y. (2017). *Early visual arts in education Hong Kong kindergartens* [Doctoral thesis]. The University of Hong Kong, Pokfulam, Hong Kong SAR.

Leung, S. K. Y. (2018). An exploratory study of early visual arts education in Hong Kong. *Journal of Research in Childhood Education, 32*, 392–403.

Leung, S. K. Y., Choi, W. Y., & Yuen, M. (2020). Video art as digital play for young children. *British Journal of Educational Technology, 51*(2), 531 554.

Li, H., Rao, N., & Tse, S. K. (2012). Adapting Western pedagogies for Chinese literacy instruction: Case studies of Hong Kong, Shenzhen, and Singapore preschools. *Early Education and Development, 23*(4), 603–621.

Livingstone, S., & Haddon, L. (2014). *EU kids online III: A thematic network to stimulate and coordinate investigation into the use of new media by children (final annual report).* IDEAS working paper series from RePEc.

Madanipour, P., & Cohrssen, C. (2019). Augmented reality as a form of digital technology in early childhood education. *Australasian Journal of Early Childhood.* https://doi.org/10.1177/1836939119885311

Maeda, J. (2013). STEM + art = STEAM. *The STEAM Journal, 1*(1), 1–3.

Marsh, J. (2012). Purposes for literacy in children's use of the online virtual world 'Club Penguin'. *Journal of Research in Reading, 37*(2), 179–195.

Marsh, J., Plowman, L., Yamada-Rice, D., Bishop, J., & Scott, F. (2016). Digital play: A new classification. *Early Years, 36*(3), 242–253.

Menzer, M. (2015). *The arts in early childhood: Social and emotional benefits of arts participation.* Washington, DC: NEA Office of Research and Analysis.

Nash, L. (2012). Exploring ipads in learning. *Learning and Teaching, 31*, 7–13.

Ng, S. S. N., & Rao, N. (2008). Mathematics teaching during the early years in Hong Kong: A reflection of constructivism with Chinese characteristics? *Early Years, 28*(2), 159–172.

Plowman, L. (2015). Researching young children's everyday uses of technology in the family home. *Open Access in Interacting with Computers, 27*(1), 36–46.

Rao, N., Ng, S. S., & Pearson, E. (2010). Preschool pedagogy: A fusion of traditional Chinese beliefs and contemporary notions of appropriate practice. In C. K. K. Chan & N. Rao (Eds.), *Revisiting the Chinese learner: Changing contexts, changing education* (pp. 255–279). Hong Kong, China: Springer.

Reitenbach, G. (2015). From STEM to STEAM education. *Power, 159*(1), 6.

Shulman, L. S. (1986). Those who understand: Knowledge growth in teaching. *Educational Researcher, 15*(2), 4–14.

Sousa, D., & Pilecki, T. (2012). *From STEM to STEAM: Using brain-compatible strategies to integrate the arts.* Thousand Oaks, CA: Corwin.

Stephen, C., & Plowman, L. (2014). Digital play. In S. Edwards, M. Blaise, & L. Brooker (Eds.), *SAGE handbook of play and learning in early childhood* (pp. 330–341). Los Angeles, CA: SAGE.

Terreni, L. (2011). Interactive whiteboards, art, and young children. *Computers in New Zealand Schools: Learning, Teaching, Technology, 23*(1), 78–100.

Vygotsky, L. (1997). Research method. In R. Rieber (Ed.), *The collected works of L. S. Vygotsky. Volume 4. The history of the development of high mental functions* (pp. 27–65). New York: Plenum Press.

Wright, S. (2007). Young children's meaning-making through drawing and "telling": Analogies to filmic textual features. *Australasian Journal of Early Childhood, 32*(4), 37–48.

Wright, S. (2014). The voice of art: Understanding children's graphic-narrative-embodied communication. In D. Machin (Ed.), *Visual communication* (pp. 515–534). Frankfurt, Germany: De Gruyter.

4

Augmented Reality in Early Childhood Education: Accessing Complex Concepts Within Imaginative Play Worlds

Rhys George and Parian Madanipour

Introduction

Examples of technology are all around us: scissors, bicycles and toothbrushes are all examples of tools that help to make work easier—they are all forms of technology. Digital technology is a particular type of technology. It uses computers as tools that make our lives easier and more efficient. Many young children have frequent access to smartphones, tablets, laptops and computers on a daily basis. Parents and early childhood educators see very young children 'swiping' at paper books in order to turn a page—evidence of how familiar they are with digital technology. Early childhood education should provide opportunities for children to explore, consolidate and extend learning, playfully, that reflect their real

The research presented as a case study in this chapter was completed in partial fulfilment of requirements for a Master of Education degree from The University of Western Australia under the supervision of Associate Professor Christine Howitt and Associate Professor Grace Oakley.

R. George (✉) • P. Madanipour
Bold Park Community School, Wembley, WA, Australia
e-mail: parian.madanipour@unimelb.edu.au

C. Cohrssen, S. Garvis (eds.), *Embedding STEAM in Early Childhood Education and Care*, https://doi.org/10.1007/978-3-030-65624-9_4

lives. Digital technology is part of their lives. Apps for mobile devices have been developed for infants, toddlers and children. As parents and early childhood professionals, carefully incorporating digital technology in early childhood curricula is important (Bird & Edwards, 2015; Department of Education, Employment and Workplace Relations, 2009; Edwards, Straker, & Oakey, 2018; Grieshaber & Yelland, 2005). Traditionally, 'technology-based learning' has typically referred to basic point-and-explore devices, such as interactive whiteboards, digital cameras and touchscreen technologies. However, new forms of technology such as augmented reality (AR) have become increasingly commonplace.

AR creates the perception that 'virtual objects'—that is, images of objects created by a digital device—are actually present in the world. At first, it may seem that AR has little to do with early childhood education. It may not even seem appropriate for early childhood at all. However, AR is increasingly commonplace in the lives of many children. It is common for children to put bunny ears or funny hats on faces whilst video-chatting with a grandparent or to create an avatar for digital messaging. These are examples of AR.

More research is needed to determine the contribution of AR to teaching and learning in early childhood (George, 2017; Madanipour & Cohrssen, 2020). However, as a teaching and learning approach, AR is attracting a lot of attention from researchers, educators and parents because it actively engages learners in the learning process. Presenting content from a three-dimensional perspective using handheld devices, desktop computers, webcams and/or head-mounted displays makes learning fun. It enables learners to visualise the invisible (Wu, Lee, Chang, & Liang, 2013). This hands-on approach to learning is why AR is so promising for helping learners to visually 'see' volcanoes, 'visit' faraway places or 'make' rain. Hence, to achieve the best learning outcome from this augmented environment, it is crucial for early childhood educators and parents to understand how children learn to use this innovative technology within a play-based learning context.

What Is Augmented Reality Technology in Early Childhood?

Books are the most obvious application of AR technology in early childhood. AR-based books supplement traditional paper books with virtual elements, such as three-dimensional virtual objects, sounds, and animations displayed over a touchscreen device. Some examples of AR books that cover the age range from two to eight years old include *Goodnight Lad*,[1] *My Very Hungry Caterpillar AR*[2] and *Rox's Secret Code*.[3] Research has found that the excitement, engagement and enjoyment which children experience when using AR technology to supplement traditional learning have a positive impact on learning (Chen & Chan, 2019; Han, Jo, Hyun, & So, 2015; Wang, Lee, & Ju, 2019; Yilmaz, Kucuk, & Goktas, 2017).

Adults play a key role in children's AR book reading. Cheng and Tsai (2014) involved parents in the learning process of AR picture book reading. Within the study, when parents and children interacted as a 'communicative child-parent pair' and children were jointly involved in the reading process with their parents, higher cognitive attainments regarding the AR book content explanation and appearance description were evident. However, low communication levels, and either the child or the parent dominating the AR book reading, were associated with low levels of observed cognitive gains. Parents also believed that learning through AR enhances a child's presence to, and interactions with, the learning material. Fostering motivation and drawing attention to reading are two further benefits that parents have raised about learning through AR (Cheng, 2017). This highlights both the importance of social scaffolding as a pedagogical approach and the increased learning efficacy AR technology can bring. Thus, the three main interactive mechanisms of traditional shared-book reading (a book, a child and an adult) can be incorporated

---

[1] Grimm, B. (2015). *Goodnight Lad | Exploring hidden worlds in books*. Retrieved from https://www.goodnightlad.com

[2] StoryToys Entertainment Ltd. (2018). *My Very Hungry Caterpillar AR*. Retrieved from https://storytoys.com/app/my-very-hungry-caterpillar/

[3] Lecocq, M. (2016). *Rox's Secret Code*. Retrieved from https://www.yoursecretcode.com/?fbclid=IwAR0D4AV5WeiM3-dHcGnIuyz-eRGH9rAzKRUUt5nS1Zpo5qMiTtEzK1RXNTw#/game

in a new innovative technology, creatively delivering an imaginary world for the readers.

Beyond books, AR mobile applications have been developed to support curricula in various domains, including science (Merge Cube[4]), engineering (Thomas & Friends Minis[5]), the arts (Quiver Education[6]), mathematics (Math Alive[7]) and literacy (Letters Alive[8]). Researchers have explored the impacts of some of these applications on children's learning. For instance, the AR application QuiverVision was found to increase motivation and participation levels in four- to five-year-old children by converting two-dimensional drawings into three-dimensional renderings. Children's enthusiasm and enjoyment throughout this AR-based art activity were indicated, particularly when seeing their drawings of aeroplanes suddenly coming alive in 3D (Huang, Li, & Fong, 2016). Higher levels of interaction and greater learning achievements with other AR applications such as Aurasma and Augment have also been portrayed (Gecu-Parmaksiz & Delialioglu, 2019; Stotz, 2018).

These findings might be due to the wonder that AR promotes in children. By overlaying a virtual layer on reality, children perceive AR as both created and real phenomena that deliver a playful environment for exploration. With no doubt, playful exploration and engagement are effective in children's learning process. For example, the preschool teachers in a recent study indicated that the use of an AR application in science prolonged the engagement and attention span of three- and four-year-old children (Ozdamli & Karagozlu, 2018). Educators generally have positive attitudes towards AR technology and believe that a hybrid learning environment which combines digital and physical objects has potential to capture children's engagement in the learning

[4] MERGE, (2020). *MERGE CUBE, the power to hold the digital world.* Retrieved from: https://mergeedu.com/merge-cube

[5] Budge Studios, (2017). *Thomas & Friends Minis.* Retrieved from: https://budgestudios.com/en/apps/detail/thomas-and-friends-minis/

[6] QuiverVision, (2016). *Quiver Education.* Retrieved from: http://www.quivervision.com/apps/quiver-education/

[7] Alive Studios Zoo, (2017). *Math Alive.* Retrieved from: https://alivestudiosco.com/math-alive-kit/

[8] Alive Studios Zoo, (2017). *Letters Alive.* Retrieved from: https://alivestudiosco.com/letters-alive-zoo-keeper-edition/

process (Chen & Chan, 2019). However, not all the perceptions Chen and Chan's (2019) study gleaned were good. The general consensus was that different learning styles must be taken into account as some children are non-visual learners and may become distracted by animation and other special effects.

The studies undertaken to date cover a wide range of curricula in early childhood, which bodes well for the strong educational potential of AR. Studies overwhelmingly show that AR technology can be used as an educational medium to improve children's understanding of the real world by enhancing it with virtual elements. Wu and colleagues. (2013) assert that seeing AR as a concept rather than a type of technology is a more constructive viewpoint for stakeholders, in that AR can and should be used as a tool. Through AR exploration, children can be provided with instant and simultaneous access to both virtual and physical information, which in turn can facilitate their learning in associating the virtual content with the real world (George, Howitt & Oakley 2019). For instance, placing a virtual aquarium right inside a classroom, and observing and exploring the scientific features of sea creatures (FishingGO AR[9]) from different perspectives, provides the learner immediate access to both worlds.

In this age of digitisation and technology, research is emphasising the role of education in developing important future-facing skills (Loble, Greenaune, & Hayes, 2017). Dunleavy, Dede and Mitchell (2009, p. 20) describe the unique ability of AR:

> to create [an] immersive hybrid learning environment that combine[s] digital and physical objects, thereby facilitating the development of process skills such as critical thinking, problem solving, and communicating utilized through interdependent collaborative exercises.

These so-called twenty-first century skills are crucial for Science, Technology, Engineering, Arts and Mathematics (STEAM) learning and so it is reasonable to infer that AR may help children to prepare for the

[9] IVANOVICH GAMES, (2017). *FishingGO AR*. Retrieved from: http://www.ivanovichgames.com/web/catalog/games/zfishinggo/

workforce requirements of the future. Early childhood educators and parents are eager to help children to develop these skills.

The Digital Play Framework

The digital play framework, developed by Bird and Edwards (2015), is a conceptual framework that explores how children learn to use technologies through play. This framework, as an assessment tool, aims to assist adults to observe and assess children's playful engagement with digital technologies. Derived from Vygotskian perspectives, the framework builds on a series of behaviours identified by Hutt, Tyler, Hutt and Christopherson (1989). It is built around the concept that children learn to use technologies through two forms of play. First, children begin to understand the functions of a given technology through exploratory or 'epistemic' play where they wonder what it does. In this form of play, children develop new skills and new knowledge (Hatzigianni, Gregoriadis, Karagiorgou, & Chatzigeorgiadou, 2018). In this framework, epistemic play involves three behaviours: (i) *exploration*, as children explore various functions of a device, (ii) *problem-solving*, when children test and evaluate different approaches on a device, and (iii) *skill acquisition*, when mastering knowledge about the function of a device happens. More advanced exploration appears when children transition into innovative or 'ludic' play. Here, children investigate and discover what they can do with that technology, demonstrating two behaviours: (i) *symbolic*, using a device to represent their thoughts symbolically, and (ii) *innovative*, where children's creative thoughts are promoted to come up with new ideas. This framework is also applied in Chap. 3 of this book.

There is a sequence that is typically followed: children's ludic play emerges after epistemic play has been mastered. However, the framework treats these two forms of play as a cycle, explaining that children might return to epistemic play to learn and develop new skills. An example of this would be a five-year-old child engaging in epistemic play while exploring the functions of a tablet's camera such as locating the shutter release button or detecting the on/off button for the video camera. The child is moving towards ludic play when he or she uses symbolic thinking to purposefully record their journey into space. The learner may go back to the first form of play while seeking assistance from a more knowledgeable person in order to view the recorded footage, or to move backwards or forwards in the video in order to watch and examine the recorded play. While mastering this skill, the child might to share learned actions with peers (Edwards & Bird, 2017).

Case Study: Augmented Reality Sandbox

The case study we present draws on the Digital Play Framework to categorise children's behaviours as they use an AR Sandbox for play. The AR Sandbox augments a traditional sandbox experience with layered, contextual visualisations.[10] A user of an AR Sandbox will see three types of visualisations on the sand surface; topographic contour lines, a colour elevation map and virtual water (see Fig. 4.1). As the user interacts with

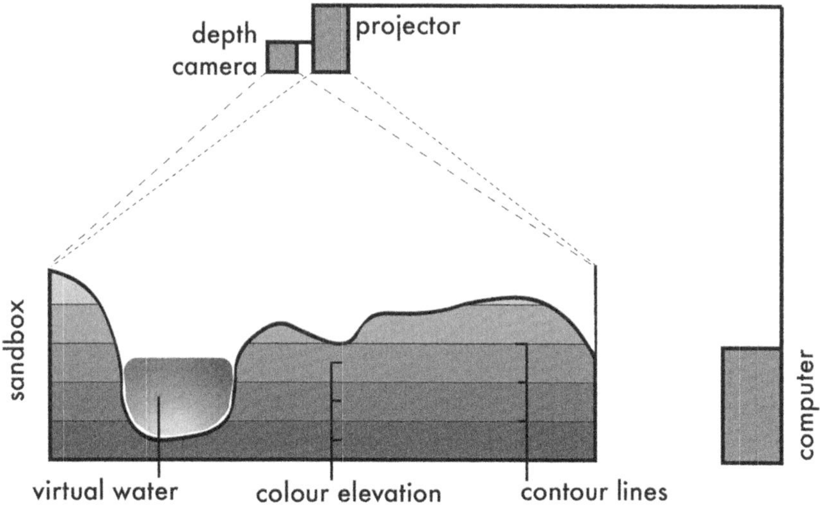

Fig. 4.1 Components of an AR Sandbox

[10] University of California, (2016). *Augmented Reality Sandbox*. Retrieved from: https://arsandbox.ucdavis.edu

the sand surface, these visualisations adapt in real time. Three main components operate these visualisations in a continuous loop. First, a camera positioned above the sandbox scans the terrain and informs a nearby connected computer. The computer then processes the scan and reconstructs an image of the sandbox. This image is received by the projector above, which vertically illuminates visualisations on the top of the sand surface. The colour elevation map lights the sand different colours depending on its relative height in the sandbox. The lowest area is coloured blue, and as the sand increases in height, it becomes green, then yellow, then red, then white. The contour lines are thin black lines that join points of equal height. The virtual water is a projected simulation that interacts with the formations within the sandbox. Multiple users can see all three visualisations simultaneously from any perspective around the sandbox.

The research was conducted at an Australian metropolitan school that adopts a social-constructivist pedagogical philosophy, inspired by Reggio Emilia. The school was deliberately chosen as their teaching practices emphasise children's learning through play. The educators are viewed as facilitators in a classroom that values children's ideas. They support children to develop their own theories and acknowledge their need to have control over their learning. The data were collected through a Mosaic approach, a model that focuses on listening to children and uses multiple ways to collect data (Clark & Moss, 2011). This approach views children as experts of their own lives and through listening, children and adults co-construct knowledge. This study focuses on a small group of four- to five-year-old children, namely Dale, Josie, Evie and Andy, within a classroom facilitated by two educators. Two mornings per week, over eight consecutive weeks, the first author (Rhys) joined the class and collected data. During each session, the researcher captured observational notes and photographs while the children interacted with the AR Sandbox and with each other. An audio recorder captured the children's discussions while they played together, and conversed with the researcher. There was a reflexive quality to the researcher's discussions with the children which assisted in clarifying the children's perspective. The children had access to nearby digital cameras which they used to capture photographs of their play. Throughout the study, the researcher and educators met regularly to share their perspectives of the children's exploration of the AR Sandbox.

All children in the classroom were invited to participate in the study through an interactive non-fiction narrative (Mayne, Howitt, & Rennie, 2017). A presentation was written to inform the children of the research aims and consent process. The children understood their participation with the research was optional and decided how they would express their consent each day (Howitt, George, & Oakley, in press). The researcher and educators reinforced that the children could play with the AR Sandbox without having to participate in the study. Approval for the study was also agreed upon by the school principal, educators, parents and the university's Human Research Ethics Committee.

The data analysis process was divided into three phases: gather and organise, code behaviours and reconstruct vignettes. First, all handwritten observations and discussions were organised into electronic documents alongside the photographs. Second, a Mosaic document was created for each of the four participant children capturing their play with the AR Sandbox over time. Third, using the Digital Play Framework, the researcher deductively reviewed the Mosaics for vignettes that would highlight children's play with the technology. The researcher looked for examples of epistemic and ludic play to demonstrate how children learn to use the AR Sandbox through play. One vignette was selected to highlight epistemic play and another for ludic play. To demonstrate the children's transition between the two play forms, a third vignette was chosen and positioned as a bridge between the two. The three vignettes were constructed using the data collected.

Findings

Epistemic Vignette: Making Shadows

This first vignette showcases the children's behaviours during their initial interactions as they play to understand the AR Sandbox's functionality. The vignette is prefaced with a table to highlight children's engagement with the technology. Table 4.1 foregrounds the children's activities as they

Table 4.1 Applying the digital play framework to understand how children learn to use the AR Sandbox through epistemic play

Behaviours	Indicators	Activities
Exploration	Seemingly random use of the AR Sandbox	Children create shadows in the projections
	Locating the operating functions of the AR Sandbox	Children observe changes in the colour visualisations when moving sand
	Exploring the operating functions of the AR Sandbox	Dale and Josie construct sand mountains
	Following directions of other people with the AR Sandbox	Dale and Josie collaboratively build a larger sand mountain
	Seeking assistance for desired outcome	Dale asks Josie about the changing colours
Problem-solving	Relating actions to the function	Josie shares her understanding of the relationship between colour visualisation and the sand height
	Trying different actions to solve an issue	Josie and Evie create shadows of different sizes to discern the projection's properties
	Intentional use of the operating functions	Dale and Josie collaboratively construct a sand mountain
Skill acquisition	Intentional and deliberate use of functions for desired outcome	Dale and Josie collaboratively construct a sand landscape
	Sharing learnt actions with others	Josie explains the impact of sand distribution as a large mountain creates a larger ocean
	Intentional and controlled use for own purpose	Children deliberately minimise their shadows to allow peers to construct within the colour visualisations

Note. Adapted from Bird & Edwards (2015)

answer the question, 'What does the AR Sandbox do?' It categorises the children's epistemic behaviours as indicated in the Digital Play Framework.

When the AR Sandbox is first introduced in the classroom, Dale and Josie head over to investigate it of their own accord. They instantly notice the visualisations and quickly begin to move the sand around. During these initial interactions, they seek understanding and share their observations of the AR Sandbox's functionality.

Dale: Why does the sand change colour?

Josie: When you dig deeper, it changes to blue, when you build up it, it changes to one of these colours.

To clarify her statement, Josie points to a sand construction illuminated green, yellow, orange and red. Dale then begins to push sand together to construct something larger. Using a nearby scoop, Josie pours sand on top of Dale's mountain as they collaborate.

Josie: Oh, wow, there is snow on Dale's mountain. Snow is coming!

As Josie pours the sand on top of the mountain, the projected light illuminates the sand a bright white. Dale and Josie continue to build the mountain together, and Josie shares her insights on the colour visualisations.

Josie: Wow! The bigger the mountain we make, the bigger the ocean is, because the more we move from the ocean, the deeper it gets, there is no more room left over.

During their construction play in the AR Sandbox, the children have been noticing the colour projections on their hands with curiosity. They move their hands beneath projections, observing how their hands are illuminated with colours, and some notice shadows forming underneath. To find the source of the bright colours, the children often look up and notice the projector's light positioned above. Josie notices the shadows forming below and deliberately begins to create shadows with her friend Evie. They move their hands both closer to the sand surface, and then further away, noticing the shadows change in size. Their curiosity develops into a science investigation, practising skills of observation and problem-solving. They each grab a piece of paper and hold the paper at different heights to compare the shadow sizes. Josie communicates her findings.

Rhys: I notice you're making the shadows bigger and smaller.
Josie: We hold the piece of paper up close to the light. The higher the piece of paper, the bigger the shadow.

Andy is observing Evie and Josie's play. He is conducting an investigation of his own. Andy notices the individual pixels of the projector on a piece of white paper. He begins counting each of the pixels when, all of a sudden, Evie uses her paper to cast a shadow over the whole sandbox.

Andy: It went to night time!

The children discuss the impact when there is no colour visualisation visible on the sand's surface. They come to a consensus that it should remain 'day time' so everyone can play in the coloured sand. They agree to stop deliberately making shadows at the AR Sandbox.

Interpretation of Making Shadows

The children's interactions in the Making Shadows vignette are evidence of curiosity, as they explored the functions of the AR Sandbox. They explored the colour visualisations by constructing sand mountains that changed the projected colours and created shadows from the projector's light. The children approached the device as a scientific investigation: observing, comparing, predicting and checking, discussing and collaborating. They shared their observations on the relationship between the sand height and the colour visualisations. Josie demonstrated her understanding of the impact of the sand distribution and colour visualisations with her peers. The children created different-sized shadows and examined the projector's properties. As they developed an understanding of the projector's properties and colour visualisations, they agreed on a rule: to minimise shadows during their play. This rule highlighted their mastery of the device by signalling a shift in their intentionality, using the AR Sandbox for its unique visualisation functionality in their play.

Table 4.2 Learning through transferring knowledge

Behaviours	Indicators	Description of activities
Problem-solving	Relating concepts from the AR functions	Scaling AR visualisations to real-world space
Skill acquisition	Sharing learnt concepts with others	Dale rationalises his scale of AR visualisations to real-world space
Symbolic	Deliberate use of AR Sandbox to represent real spaces	Dale uses visualisations to create a model of nearby lake

Note. Adapted from Bird & Edwards (2015)

Transitional Vignette: Lake Packard

The second vignette showcases a transition from epistemic play to a ludic phase of play and development in understanding the technology. Table 4.2 highlights how Dale answers, 'What can I do with the AR Sandbox?' and critically reflects on 'What does the AR Sandbox do?'

In the second week of the installation, Dale proposes a play idea at the AR Sandbox.

Dale: Hey guys, let's make it Lake Packard in here.
Rhys: How are we going to make Lake Packard in here?
Dale: Well, it goes straight and then it curves, then it goes straight, then it goes curved there, and then it's straight and curves back around.

Dale brings a selection of blocks over to the sandbox to create a perimeter for his design. He digs out the sand inside his edge of the sandbox, manipulating the visualisations to turn blue. Dale takes a photograph of his construction and explains, 'This is Lake Packard'.

The following week, the class walk to the nearby Lake Packard where they visit a hill that they enjoy running up and down together. Once finished, the teacher asks, "If the hill was in the sandbox, what colour do you think it would be?" The excited class offers a range of suggestions. All the colours of the sandbox are mentioned except blue. Dale is among the children asked to share his opinion with the class, to which he responds,

"Green, yellow, red and white." As the class walks back to school, Rhys approaches Dale asking him to reflect on why he chose those colours.

Dale: I have had a think and the hill would be green because the hill was actually green.
Rhys: So it depends on the *actual* colour of the hill?
Dale: Hills are smaller than mountains and mountains are bigger than hills. Green is land and it has to be above the water. It's like a bathtub, the water goes to the bottom and stays there.

Interpretation of Lake Packard

The Lake Packard vignette demonstrated how Dale moved between the two forms of play (epistemic and ludic). Dale built a model of a real-world space, deliberately utilising the colour elevation map, unique to the AR Sandbox. He dug out the middle to represent the lake as blue. When the children visited Lake Packard, the concepts experienced in the AR Sandbox were introduced into a real-world environment. When Dale reflected on the educator's question, he had to problem-solve how to contextualise the concepts from the AR Sandbox in the real world. He thought critically about what colour the real-world space was and shared his response. Dale reasoned the hill was green as it symbolically represented the grassy environment and it also provided a relative height scale. The adult promotes the thinking of the child through reflection of his response. Dale's understanding of the AR Sandbox visualisations develops. His relative colour elevation scale continues in the following vignette, as it is implemented in the children's creation of Dinosaur Park.

Ludic Vignette: Dinosaur Park

The Dinosaur Park vignette showcases how the children build upon their mastery of the AR Sandbox's functionality into a new imaginative play experience. Table 4.3 introduces the children's play as they ask the question, 'What can I do with the AR Sandbox?'

Table 4.3 Applying the digital play framework to understand how children learn to use the AR Sandbox through ludic play

Behaviours	Indicators	Description of activities
Symbolic	Seamless use of real objects and AR Sandbox for pretend play	Children use a mix of virtual and real objects in construction of Dinosaur Park
	Deliberate use of AR Sandbox for pretend play	Children use AR colour visualisations in their dinosaur play
Innovation	Purposeful pretend play created for the use of AR functions	Children enable and integrate virtual water visualisation in construction of Dinosaur Park
	Deliberate manipulation of AR functions for pretend play	Children manipulate the virtual water for pretend play with dinosaurs

Note. Adapted from Bird & Edwards (2015)

In the fifth week of the study, the children were incorporating dinosaur figurines in the AR Sandbox.

Dale: I started making a volcano then I changed it into a Dinosaur Park. This is the dry land and they would get really thirsty.
Andy: The dinosaur needs to eat trees to survive.
Dale: I want to go outside and get some trees for the dinosaur to live.

During playtime, the children have access to an area outside the classroom filled with natural materials. Dale and Andy visit the space and return with sticks and leaves.

Josie: I have lots of dinosaurs…they are in the water having a drink at the waterhole.
Andy: My dinosaur is on the fence; he is about to jump.
Dale: Don't let him escape! The T-Rex is going to eat all the trees.
Andy: The T-Rex are going to eat the other dinosaurs…the T-Rex is digging at the fence.
Dale: He just wants to explore.

Josie, Andy and Dale are utilising the colour visualisations as they create walls and pathways between sections in their Dinosaur Park. Dale captures a photograph of his part of the park to show how the dinosaurs live on the green areas. Like Josie's waterhole, Dale has been taking his dinosaurs for a drink where the green meets the lower blue areas. His section is surrounded by large walls that are illuminated red, intended to keep the dinosaurs in. The dinosaurs in Andy's section are trying to escape.

Andy: Mine keeps getting water from this side, what about I could join my pool to your pool?
Dale: Yeah and then the dinosaurs could come down there and we could visit.
Andy: Then we could make a stream. They can visit through the stream.
Dale: I could make a dry path from here and then you can make a wet path.

After Andy and Dale discuss pools, streams and paths, Dale requests the virtual water feature be enabled. Josie and Andy agree to allow the function and the children create rain that interacts with their sand constructions.

Dale: It rained down and fills up the holes and gets full and drip drops off the side like rain.
Josie: I've got a waterhole, and this is where they use the water for a swimming pool.
Dale: I have built a bridge for the dinosaurs to cross (Fig. 4.2).
Josie: The sharks are in the water to protect the treasure by the leaf.

The children sculpt the landscape to incorporate the streams and pathways that direct the virtual water around their Dinosaur Park. The dinosaurs begin to move up to the higher ground as the water pathways become the boundaries for the children's sections. Josie's shallow swimming pool has virtual water within it. Dale experiments constructing bridges that allow his dinosaurs to join Andy's area. He uses sand to create his first bridge, adapting a technique the children had begun during the

Fig. 4.2 Dale builds his second bridge over the virtual water using sticks from the outside area

second week of the study. Further along the stream, Dale constructs his second bridge using the sticks he had previously used as trees that the dinosaurs ate.

Interpretation of Dinosaur Park Vignette

Throughout the vignette, the children's symbolic use of the dinosaur figurines and sticks seamlessly merges with their symbolic use of the AR digital visualisations. Dale and the other children build upon their experience from the Lake Packard vignette, where the colour visualisations were contextualised to a real-world landscape. In their Dinosaur Park, the dinosaurs lived in the green areas and were enclosed by the red walls that were too high to climb. As their collaborative pretend play developed, they explored engineering concepts by creating bridges for the dinosaurs to use. The children adapted the landscape of Dinosaur Park for the virtual water functionality. They creatively sculpted streams, pools and pathways in a way that channelled the virtual water for their dinosaurs to drink and partitioned sections of their park.

As they included the virtual water in their play, it increased the complexity of the use of AR in their shared world. The children built upon their understandings to purposefully integrate the AR visualisations to transform their play. They create a shared an imaginary world that could only exist with the AR Sandbox's unique functions.

Discussion

With the use of the Digital Play Framework, this study delivers useful insights into how children learn to use the AR Sandbox through play. Children built upon their understandings through epistemic play to ludic play. In the first form of play, children engaged in playful exploration activities to investigate and understand the functions of the new digital technology, the AR Sandbox. In the second form of play, children progressed to a more imaginative use of the AR features by using symbolic and innovative thought. Within these two forms of play, the AR Sandbox demonstrated its ability to move beyond a traditional sandbox to provide a hybrid learning environment that facilitated learners' development of twenty-first century skills by extending the complexity of their play. Critical thinking, problem-solving, creativity, collaboration and communication are the key examples of twenty-first century skills that children showed while interacting with the AR Sandbox.

Within this context, the Digital Play Framework may be a suitable assessment tool for both parents and early childhood educators in order to understand children's digital play. It is important to note the significant role played by adults to promote and facilitate children's thinking and understanding through interacting with the AR. Within the delivered study, the interactions among the adults and the learners in developing knowledge and meaning together assisted the participants to share and clarify their understanding of the AR Sandbox through both forms of play. The adult and the children contributed to the thinking to develop and extend their understanding of the AR Sandbox. This demonstrates the crucial contribution made by early childhood educators and parents, knowing how children learn to use AR technology through play, when they involve themselves in children's digital play to clarify concepts and evaluate the learning experiences collaboratively.

Conclusion

One of the key pedagogical affordances of AR is its ability to create immersive, hybrid learning environments that combine digital and non-digital elements. Users can experience both the virtual and the real in a way that gives new meaning to both. The AR Sandbox extends the child-centred play experience of a traditional non-digital sandbox by providing augmented experiences accessible to all ages. For example, infants and toddlers could simultaneously access digital functions in a way that does not interfere with their tangible experiences. As a digital technology for learning in early childhood, the AR Sandbox provides opportunities to explore abstract concepts contextualised within the world of children's collaborative play. Thus, AR can be used to create a rich learning environment suitable for early childhood as it allows children to develop twenty-first century skills during play. The AR Sandbox installation is only one such example. Consequently, more experimental research is required to examine the feasibility of other forms of AR in early childhood contexts.

When given time to play with technology, children—and indeed adults—deepen their understanding and internalise the concepts that are fundamental to its use. This in turn encourages the user to think in new ways. As humans create new technologies, they build upon the cumulative cultural knowledge and skills of those before them. Thus, new technologies like AR influence both the thinking of the individuals using it, and culture itself.

By definition, AR adds a layer of information to a user's experience. Conceptually, it shapes a user's perspective, allowing abstract concepts to merge with concrete experiences. The children in this case study used a complex framework of thinking within a play world they had constructed. This is one of the greatest affordances of AR for young children's learning: it allows children to access complex concepts within their imaginative play.

References

Bird, J., & Edwards, S. (2015). Children learning to use technologies through play: A digital play framework. *British Journal of Educational Technology, 46*(6), 1149–1160. https://doi.org/10.1111/bjet.12191

Chen, R. W., & Chan, K. K. (2019). Using augmented reality flashcards to learn vocabularies in early childhood education. *Journal of Educational Computing Research, 57*(7), 1812–1831. https://doi.org/10.1177/0735633119854028

Cheng, K.-H. (2017). Exploring parents' conceptions of augmented reality learning and approaches to learning by augmented reality with their children. *Journal of Educational Computing Research, 55*(6), 820–843.

Cheng, K.-H., & Tsai, C.-C. (2014). Children and parents' reading of an augmented reality picture book: Analyses of behavioral patterns and cognitive attainment. *Computers & Education, 72*, 302–312. https://doi.org/10.1016/j.compedu.2013.12.003

Clark, A., & Moss, P. (2011). *Listening to young children: The mosaic approach* (2nd ed.). London: National Children's Bureau.

Department of Education, Employment and Work-place Relations (DEWWR). (2009). *Belonging, being & becoming: The early years learning framework for Australia*. Canberra, ACT: Commonwealth of Australia.

Dunleavy, M., Dede, C., & Mitchell, R. (2009). Affordances and limitations of immersive participatory augmented reality simulations for teaching and learning. *Journal of Science Education & Technology, 18*(1), 7–22. https://doi.org/10.1007/s10956-9119-1

Edwards, S., & Bird, J. (2017). Observing and assessing young children's digital play in the early years: Using the digital play framework. *Journal of Early Childhood Research, 15*(2), 158–173.

Edwards, S., Straker, L., & Oakey, H. (2018). *Statement on young children and digital technologies*. Retrieved from http://www.earlychildhoodaustralia.org.au/our-work/submissions-statements/eca-statement-young-children-digital-technologies/

Gecu-Parmaksiz, Z., & Delialioglu, O. (2019). Augmented reality-based virtual manipulatives versus physical manipulatives for teaching geometric shapes to preschool children. *British Journal of Educational Technology, 50*(6), 3376–3390.

George, R. (2017). *Young children's use of an augmented reality sandbox for spatial thinking through play* (Master's thesis). Retrieved from https://research-repository.uwa.edu.au/en/publications/young-childrens-use-of-an-augmented-reality-sandbox-for-spatial-t

George, R., Howitt, C., & Oakley, G. (2019). Young children's use of an augmented reality sandbox to enhance spatial thinking. *Children's Geographies.* https://doi.org/10.1080/14733285.2019.1614533

Grieshaber, S., & Yelland, N. (2005). Living in liminal times: Early childhood education and young children in the global/local information society. In M. Apple, J. Kenway, & M. Singh (Eds.), *Globalizing education: Policies, pedagogies and politics* (pp. 191–207). New York: Peter Lang.

Han, J., Jo, M., Hyun, E., & So, H. J. (2015). Examining young children's perception toward augmented reality-infused dramatic play. *Education Tech Research Dev, 63*, 455–474. https://doi.org/10.1007/s11423-015-9374-9

Hatzigianni, M., Gregoriadis, A., Karagiorgou, I., & Chatzigeorgiadou, S. (2018). Using tablets in free play: The implementation of the digital play framework in Greece. *British Journal of Educational Technology, 49*(5), 928–942.

Howitt, C., George, R., & Oakley, G. (in press). Using a narrative approach informing story to engage 4- and 5-year olds in research in the context of a study on the use of an augmented reality sandbox to enhance spatial thinking. In F. Mayne & C. Howitt (Eds.), *The narrative approach and informing story: making young children's research participation meaningful.* Routledge.

Huang, Y., Li, H., & Fong, R. (2016). Using augmented reality in early art education: A case study in Hong Kong kindergarten. *Early Child Development and Care, 186*(6), 879–894. https://doi.org/10.1080/03004430.2015.1067888

Hutt, S., Tyler, C., Hutt, C., & Christopherson, H. (1989). *Play, exploration and learning. A natural history of the preschool.* London: Routledge.

Loble, L., Greenaune, T., & Hayes, J. (2017). *Future frontiers: Education for an AI world.* Melbourne, Australia: Melbourne University Press.

Madanipour, P., & Cohrssen, C. (2020). Augmented reality as a form of digital technology in early childhood education. *Australasian Journal of Early Childhood, 45*(1), 5–13. https://doi.org/10.1177/1836939119885311

Mayne, F., Howitt, C., & Rennie, L. (2017). Using interactive nonfiction narrative to enhance competence in the informed consent process with 3-year-old children. *International Journal of Inclusive Education, 21*(3), 299–315. https://doi.org/10.1080/13603116.2016.1260833

Ozdamli, F., & Karagozlu, D. (2018). Preschool teachers' opinion on the use of augmented reality application in preschool science education. *Croatian Journal of Education, 20*(1), 43–74. https://doi.org/10.15516/cje.v20i1.2626

Stotz, M. D. (2018). *Creature counting: The effects of augmented reality on perseverance and early numeracy skills* (Doctoral dissertation). Retrieved from https://preserve.lehigh.edu/cgi/viewcontent.cgi?article=5255&context=etd

Wang, L., Lee, H., & Ju, D. Y. (2019). Impact of digital content on young children's reading interest and concentration for books. *Behaviour & Information Technology, 38*(1), 1–8. https://doi.org/10.1080/0144929X.2018.1502807

Wu, H.-K., Lee, S. W.-Y., Chang, H.-Y., & Liang, J.-C. (2013). Current status, opportunities and challenges of augmented reality in education. *Computers & Education, 62*, 41–49. https://doi.org/10.1016/j.compedu.2012.10.024

Yilmaz, R. M., Kucuk, S., & Goktas, Y. (2017). Are augmented reality picture books magic or real for preschool children aged five to six? *British Journal of Educational Technology, 48*(3), 824–841. https://doi.org/10.1111/bjet.12452

5

Screen-Free STEAM: Low-Cost and Hands-on Approaches to Teaching Coding and Engineering to Young Children

Amanda Sullivan and Amanda Strawhacker

Introduction

In the United States and worldwide, there has been a growing focus on promoting Science, Technology, Engineering, and Mathematics (STEM) education during the early childhood and elementary years (National Science and Technology Council, 2018). This may be due in part to the noticeable lack of professionals qualified to take on jobs in the sciences. In less than a decade from now, it is estimated that the United States will need 1.7 million more engineers and computing professionals (Corbett & Hill, 2015). Early childhood and early elementary school is a critical time to reach future scientists and engineers in order to meet this growing workforce need (Bers, 2012, 2018; Sullivan, 2019). Children who are exposed to STEM curricula and programming at an early age

A. Sullivan (✉) • A. Strawhacker
Tufts University, Medford, MA, USA
e-mail: amanda.sullivan@tufts.edu

C. Cohrssen, S. Garvis (eds.), *Embedding STEAM in Early Childhood Education and Care*, https://doi.org/10.1007/978-3-030-65624-9_5

demonstrate fewer gender-based stereotypes regarding STEM careers, an increased interest in engineering, and fewer obstacles entering these fields later in life (Madill et al., 2007; Markert, 1996; Metz, 2007; Steele, 1997; Sullivan, 2019; Sullivan & Bers, 2017). Moreover, we have seen many cognitive and social benefits of implementing STEM, and particularly computer science, robotics, and engineering curricula with young children (e.g. Bers, 2008; Fessakis, Gouli, & Mavroudi, 2013; Kazakoff, Sullivan, & Bers, 2013; Lee, Sullivan, & Bers, 2013).

Despite the research, actually *reaching* children with quality STEM content, particularly with regard to the "T" of technology and "E" of engineering, during their foundational early childhood years has proven to be a real challenge to many parents and educators. Choosing developmentally appropriate ways to address fields like engineering and computer science with young children presents both practical and ethical issues for adults to consider. Innovative technologies to support STEM learning such as iPads, robotics kits, and computers are expensive, and often the cost of these materials (let alone the cost of training and professional development for adults on how to use them effectively) makes them out of reach for many parents and educators. This has opened the door to a new type of "digital divide" in which some schools and homes have access to high-quality STEM and computing devices while others do not.

In this chapter, we present a different approach to exploring technology, engineering, and the sciences during the early childhood years. By focusing on screen-free, low-tech, and collaborative approaches to topics such as engineering and coding, we demonstrate that it is possible to teach and learn technical STEM skills without access to expensive digital technology and kits. These inclusive activities are designed to be accessible to children of any gender and background within the age range of approximately three to eight years of age, and can be implemented in both home and school settings. The activity examples within this chapter highlight a Science, Technology, Engineering, Arts, and Mathematics (STEAM) rather than STEM approach to designing curriculum. By integrating the arts with the sciences, this chapter explores the ways that domains such as computer programming and engineering can be enhanced by infusing opportunities for creativity and artistic expression as well as an integration with other early childhood curricular content.

STEAM in Early Childhood Education

The STEAM Movement

Early childhood STEM education has historically focused on building foundational numeracy skills and an understanding of the natural sciences for young children (Bers, 2008; Bers, Seddighin, & Sullivan, 2013; Moomaw & Davis, 2010). In the growing national and international level discussion around STEM, *how* to effectively teach technology and engineering has become more pressing to researchers and educators (National Science and Technology Council [US], 2018; UK Department of Education, 2013; US Department of Education, 2010). This concept of promoting creativity and expression through technology and science is articulated in a newer acronym called "STEAM" (Science, Technology, Engineering, Arts, Mathematics) that is growing in popularity across the United States and worldwide (Allen-Handy, Ifill, Schaar, Rogers, & Woodard, 2020; Watson, 2020; Yakman, 2008). The "A" of STEAM can represent more than just the visual arts, but also the liberal arts, language arts, social studies, music, and more.

Within an early childhood context, STEAM education means finding ways for children to explore these subjects in an integrated way through hands-on projects, books, discussions, experiments, art explorations, collaboration, games, physical play, and more. New technological tools such as programmable robotics kits and programming languages designed for young children have become a popular way to teach interdisciplinary STEAM content by integrating arts and crafts, literacy, music, and more with engineering and robotics (Barnes, FakhrHosseini, Vasey, Park, & Jeon, 2020; Bravo Sánchez, González Correal, & Guerrero, 2017; Elkin, Sullivan, & Bers, 2016; Sullivan, Strawhacker, & Bers, 2017). Robotics kits have evolved in the tradition of educational manipulatives that allow children to explore their understanding of shape and number, spatial relations, and proportion (Brosterman, 1997; Kuh, 2014; Nicholson, 1972; Resnick et al., 1998).

In research trials with simple robotics and programming languages, children as young as four years old have demonstrated understanding of foundational engineering, programming, and robotics content (Bers, Ponte, Juelich, Viera, & Schenker, 2002; Cejka, Rogers, & Portsmore, 2006; Sullivan, Kazakoff, & Bers, 2013; Sullivan & Bers, 2015; Perlman, 1976; Wyeth, 2008; Zviel-Girshin, Luria, & Shaham, 2020). In addition to mastering this new content, programming interventions have been shown to have positive benefits for children's developing numeracy, literacy, and visual memory, and can also prompt collaboration and teamwork (Clements, 1999; Lee et al., 2013). Moreover, we have seen young children use robotics kits to explore more than engineering and coding, including culture, dance, music, and more within an integrative STEAM context (e.g. Kim & Kim, 2020; Sullivan & Bers, 2017).

While there is growing evidence that programming education supports children's attitudes and interest in STEM fields, research is ongoing about the cognitive benefits of learning to code (Rodriguez, Rader, & Camp, 2016). Critics argue that it is unclear whether or how the knowledge that learners acquire when programming (often called computational thinking) can transfer to contexts beyond the coding environment (e.g. Greiff et al., 2014; Scherer, 2016). In a recent meta-analysis of transfer in computer programming education, the authors found a moderate overall transfer effect between computer science learning and other cognitive skills such as creativity, reasoning, mathematics, and metacognition (Scherer, Siddiq, & Sánchez Viveros, 2019). One conclusion from this work is that learners show high ability to apply programming knowledge in similar contexts to their learning environment, such as completing a novel task using a familiar programming platform (Scherer et al., 2019). This finding has yet to be confirmed in non-technological contexts, such as when children engage in "unplugged" (non-technological) coding activities (Hickmott, Prieto-Rodriguez, & Holmes, 2018). However, preliminary studies of the comparative effect of "unplugged" and technology-based coding activities on computational thinking found no differences between children who completed unplugged and those who used tablet-based coding activities (Messer, Thomas, Holliman, & Kucirkova, 2018; Rodriguez et al., 2016). Indeed, one study found that children who completed unplugged coding activities showed significantly higher computational thinking than a non-coding control group (Brackmann et al., 2017).

Although research is ongoing as to the cognitive outcomes of programming and robotics knowledge to different settings, researchers do tend to agree that unplugged coding and engineering activities are a useful way for children and adults to meaningful and positively engage with novel STEAM domains (e.g. Bell & Vahrenhold, 2018). Research on early childhood STEM education confirms that parents and teachers are critical for supporting children's positive early STEM experiences, but that they need training and resources to effectively foster STEM learning (Bell & Vahrenhold, 2018; McClure et al., 2017; Strawhacker, Lee, & Bers, 2017). This poses a challenge since teachers may not have had professional STEM training, but studies show that professional development experiences that teachers who used unplugged, story-based, and physical STEAM activities, like the ones we present in this chapter, expressed confidence and willingness to integrate STEAM domains into their classroom settings (Bell & Vahrenhold, 2018; Curzon, McOwan, Plant, & Meagher, 2014; Sentance & Csizmadia, 2017; Smith et al., 2015).

STEAM and the New Digital Divide

There are now many digital tools, such as the robotics kits previously mentioned, available for young children to explore STEAM. But many of these new tools, despite their benefits, are inaccessible due to the cost, technical support, and professional development needed to implement them properly. For example, the KIBO Robotics Kit, developed by KinderLab Robotics for children aged four to seven years, offers a screen-free and hands-on kit that has decades of research highlighting its educational benefits (e.g. Sullivan & Bers, 2015; Sullivan, Bers, & Mihm, 2017). But with a cost of \$220–\$500+ per kit, it is unfortunately beyond the budget of many early childhood educators and parents. Similarly, the LEGO WeDo robotics kit for children seven years and older costs \$221 per kit and requires the use of a tablet or other device for programming. Bee-Bot, one of the cheaper robots for young children, is still around \$60 per robot, without any other accessories, and without allowing for the building and constructing components that KIBO and LEGO WeDo offer. Even free coding applications and games require schools or homes to have access to tablets and computers for each child for them to be used as intended. The costs of these devices alone are already prohibitive to many, without factoring in fees and time for training and professional development for adults to feel confident using these tools with young children.

The stark costs of new coding and engineering materials for young children have opened the door to a new type of digital divide. The term "digital divide" once simply referred to whether classrooms or homes had computers and Internet access. Now that most homes and schools have Internet connectivity basic hardware, this phrase has taken on a new meaning. There is now a socioeconomic division between those with access to high-quality, open-ended software and technology that promotes creative STEAM learning and those who do not. For example, access to computer science classes and clubs is generally lowest for students from lower-income households (Busteed & Sorenson, 2015). Inequitable access to computer science education could place these students at a disadvantage as computer technology continues to advance, especially as coding is thought of as "the new literacy" in this day and age (Bers, 2018).

Low-Cost and Screen-Free Materials and Activities

Digital technology, games, robotics kits, and more can be wonderful ways to explore STEAM at the early childhood level (see Sullivan, Strawhacker, & Bers, 2017 for ideas on using robotics within a STEAM context). In this chapter, we simply hope to demonstrate that expensive technology is not the *only* way to teach coding and engineering to young children. In an attempt to reach all young children, we focus on presenting STEAM activity ideas and materials that are low-cost and accessible to all, in order to help bridge the divide in access to computer science and engineering education. All the activities and approaches can be done in homes or classrooms that are strictly technology and screen-time free, or they can be used to supplement other curricula that use computers, tablets, programmable robotics kits, and more. We will begin by exploring STEAM materials for toddlers to explore engineering and progress to materials, resources, and approaches for teaching computer science and engineering to children in Kindergarten through second grade.

Exploring STEAM with Toddlers

A screen-free and hands-on approach to exploring STEAM may be especially useful for those parents and educators working with young children under the age of four years. The American Academy of Pediatrics recommends that preschool-aged children, between the ages of two and five, should have limited screen-time each day (American Academy of Pediatrics, 2018). Therefore, STEAM exploration for very young children should focus on providing them with multisensory, hands-on experiences that engage their senses and build on their natural curiosity.

It is typically the "T" in STEAM that gives adults pause when thinking about reaching very young children. It is important to remember that technology does not just have to mean expensive electronic devices, computers, and tablets. Rather, we can think of technology simply as any human-made tool that allows us to solve a problem or complete a task more easily. When it comes to very young children, some developmentally appropriate tools to explore may include child-safe scissors, tongs, eye-droppers, magnifying glasses, ramps, and more. Toddlers can explore engineering and mathematics through building and experimenting with blocks, puzzles, building bricks, magnetic tiles, and more. Asking children questions and encouraging them to make hypotheses and observations while they play can help foster scientific inquiry and an engineering mindset.

Activity Example

In this example, a classic building activity from early childhood is augmented with prompts developed and tested by the authors during educational STEAM research interventions at schools, makerspaces, and weekend workshops with children. When exploring engineering and design with young children, building tall towers is one of the easiest activities parents and educators can implement (See Fig. 5.1). Young children naturally explore stacking with blocks and nesting cups when they are very young. By the time they reach preschool, many children are very adept at stacking (and knocking down!) structures. For this simple activity, almost any materials you have available can work, from blocks and building bricks to recycled materials like plastic cups, paper towel rolls, and more.

Fig. 5.1 Toddler-created tower built with magnetic tiles

This is an easy "free play" activity for children to explore on their own. Adults should focus on asking prompting questions that focus children on engineering and scientific method concepts. For example, adults could take this activity deeper through one or more of the following:

- Provide children with a variety of different tower-making materials. Encourage children to predict, or guess, which material will allow them to build the tallest tower.
- Ask children to think about what makes their tower sturdy, or strong. Is it having a wider base? Is it using a particular material?
- Help children to measure their towers and record the measurements.
- Encourage children to test and improve their towers. What is one thing children could change about their designs to make them more functional (i.e. stronger, taller, wider, etc.)?

While this activity has an explicit focus on science, mathematics, and engineering, it can easily integrate into a longer STEAM curriculum unit as well. For example, children can focus on the art and design of their towers by allowing them time to work with paints, crayons, or other craft materials that allow them to decorate and express their creativity. Children could also move on from building towers to building replicas of their own neighborhoods including houses, schools, supermarkets, and other neighborhood landmarks. This could be one part of a larger interdisciplinary social studies unit that focuses on community and mapping, but also on art and engineering, as children create and decorate community maps for their structures to sit upon.

Computer Science Unplugged for Young Children

For toddlers, we have seen that the "T" of technology in STEAM can focus on simple human-made tools like pencils, scissors, and more. The previous section focused on fostering engineering within an interdisciplinary STEAM context, rather than on technical areas like computer science. As young children grow older, they become more curious about how other elements in their human-made world around them work.

They wonder how things like cell phones and computers function. This becomes an opportunity to teach young children about technology and computer science. Children can learn that their favorite apps and digital games all work because of code. They can learn that just like they are learning to read and write in English, Spanish, or any other language, and they can also learn to read, write, and create code!

While there are many benefits to teaching coding to young children, the complications of screen-time and reliance on expensive devices present roadblocks in terms of accessibility. The "unplugged" approach to computer science education has become a powerful movement over the past two decades, as educators have recognized the value of integrating activities that do not require knowledge of computers or other technologies into the computer science curriculum (Bell & Vahrenhold, 2018). This unplugged (i.e. "tech free") approach focuses on teaching programming concepts through puzzles, games, art, and more, all without a computer, robot, or tablet. Unplugged approaches to computer science claim to enable the development of computational thinking without spending time or cognitive resources on syntax and grammar of programming languages (Bell, Alexander, Freeman, & Grimley, 2009; Bell, Witten, & Fellows, 1998). The original Computer Science Unplugged project was based at Canterbury University and has since been widely adopted internationally (translated into 12 languages), and it is also recommended in The Association of Computing Machinery (ACM) K-12 curriculum (Bell et al., 2009).

Computer Science Unplugged Activity Example

This example comes from free resources posted on the CS Unplugged website by the Computer Science Education Research Group at the University of Canterbury (the authors share no affiliation with this research group) (Bell et al., 2009). By visiting csunplugged.com, parents and educators can find a range of unplugged activities to implement with young children. The website has activities organized by topic and age range (see Fig. 5.2). For example, there is a list of activities to explore binary numbers, error detection, searching algorithms, and more. The

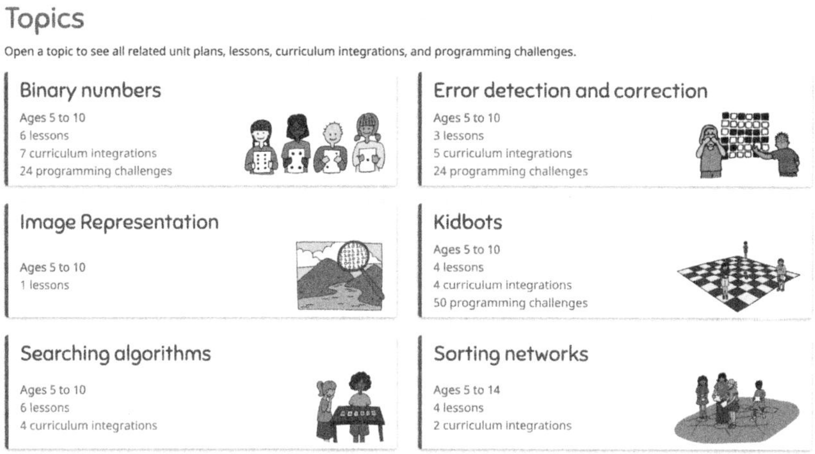

Fig. 5.2 Screenshot of activity topics on CS Unplugged website

prompts suggested for each topic typically involve a hands-on activity that may include the use of arts and crafts or other tangible materials, group discussion prompts, and ideas for play and exploration.

One lesson example presented for children as young as five years of age involves the binary number system. Why is the binary number system important for us to know about? Binary code is how computers talk and represent information. Children can think of binary as a fun new number language to explore. Children may be interested to learn that letters, numbers, and pictures (basically everything you see on the computer) is made up of different combinations of 0's and 1's.

Binary is a base-2 number system. This sounds complicated, but is just a bit different from the more common decimal, or base-10, number system. Every number "place" in our base-10 system is a multiple of 10, and we combine 10 digits (0–9) to create any number we want. For example, the number 158 only uses only three digits, but the *order* of the numbers matter: there is a 1 in the hundreds-place, a 5 in the tens-place, and an 8 in the ones-place. In binary, the system is exactly the same except that there are only 2 digits (0 and 1), and all the number places are multiples of 2. Computers use binary because it is simpler for a machine to understand than the complex decimal system. A 5-digit binary system can express numbers from 0–99,999. This is called "5 bit," and it actually is a shortened phrase that simply means "5 binary digits." CS Unplugged provides a helpful 5-bit binary to alphabet key that adults can adapt into posters or worksheets for their students (see Fig. 5.3). In 5-bit binary, each English letter can be represented by a combination of five 0s and 1s.

In the CS Unplugged activity, children create a necklace with their initials written in 5-bit binary. Adults do not need to cover the whole of binary to run this activity. Instead, this project is simply intended to be a fun and hands-on introduction to how computers store information. To complete this activity, decide which bead color will represent 1 and which bead will represent 0. For example, 0 could be blue and 1 could be red. Next, children choose their letters and see how their initials are translated into binary and then into colored beads. For example, the letter A (00001 in binary, see Fig. 5.3) would be represented by the following beads: blue, blue, blue, blue, red. To make this activity even simpler for young children, adults can create a poster showing a direct translation of

Base 10	Binary	Letter	Base 10	Binary	Letter
0	00000		14	01110	n
1	00001	a	15	01111	o
2	00010	b	16	10000	p
3	00011	c	17	10001	q
4	00100	d	18	10010	r
5	00101	e	19	10011	s
6	00110	f	20	10100	t
7	00111	g	21	10101	u
8	01000	h	22	10110	v
9	01001	i	23	10111	w
10	01010	j	24	11000	x
11	01011	k	25	11001	y
12	01100	l	26	11010	z
13	01101	m			

Fig. 5.3 Screenshot of alphabet to 5-bit binary key from CS Unplugged

how each letter of the alphabet would be represented by colored beads as well as worksheets for children to lay their beads on before stringing them.

In addition to exploring the concept of binary, activities like this can easily integrate into a longer STEAM curriculum that integrates art, fashion, and design in the creation of all sorts of jewelry, the creation of friendship bracelets, and more. While a deep understanding of binary is not needed for this activity, a basic understanding of what binary is can be helpful. CS Unplugged also provides a 45-minute lesson plan that can be used in conjunction with this activity that includes guided prompts on how to first introduce the topic of binary number system to young children, before getting into this hands-on activity.

Board Games to Explore Coding

From within the CS Unplugged movement, a new crop of unplugged coding board games and card games has been growing in popularity over the past decade. Board games and card games are some of the easiest (and most fun!) ways to explore computer programming with young children because you do not need a computer or any other device. Coding board games are also more conducive to learning and playing in home environments and informal education environments because they can be played with multiple players of mixed ages. Playing these games as a family can help younger children learn and understand the rules of the games faster than if they were to play by themselves. Table 5.1 outlines a few examples of popular coding board games that are designed to reach players younger than eight years of age. All of these examples are available for less than US$25, making board games a cost-effective solution for those without access to expensive tablets, computers, or robotics kits.

In addition to these coding-explicit games, parents and educators should remember that many traditional board games like Chess, Go and Backgammon can also be used to teach and reinforce the same problem-solving and strategy skills that are necessary across STEAM disciplines. Board games also help to teach young children important interpersonal skills such as patience, turn-taking, and being a gracious winner/loser.

Table 5.1 Coding board games for young children

Board game	Age range	Cost and description	STEAM skills
Robot Turtles	3+	$21—multiplayer board game with the goal of programming your turtle to navigate a maze to reach its jewel.	Sequencing Problem-solving Debugging Functions Planning Turn-taking
LittleCodr	4+	$13—a card game in which children program their parents or friends to do crazy things by using simple action cards	Logic Sequential thinking Prototyping debugging Turn-taking
Coding Farmers	7+	$14—children play the game with action cards in two ways: regular English and Java code. By playing the game several times, children learn to connect their actions with written code.	Java programming Addition Subtraction Reading/ vocabulary Turn-taking

Note. All prices given in USD

Board Game Example: Robot Turtles

This example comes from a board game developed by a software engineer who wanted a way to teach coding to his young children and was produced by a private company called ThinkFun (www.thinkfun.com) (the authors share no affiliation with the developers or producers of Robot Turtles) (Shapiro, 2014). The Robot Turtles board game teaches coding concepts to children ages three years and older, and was the most-backed board game in Kickstarter history at the time its campaign closed in 2014 (Shapiro, 2014). The game setup and rules of Robot Turtles are easy for young children to master: you create a maze on the board with the turtles in the corners and the jewels in the center (see Fig. 5.4). Children play instruction cards during their turn (such as, turn right, turn left, move forward, etc.) in order to "program" their turtles to get to their jewels. When a player's turtle reaches their jewel, they win. If they make a mistake, they can use a "Bug" card to undo a move. Creating these programs

Fig. 5.4 Robot Turtles board game

encourages sequential thinking and problem-solving: two key components of computer programming.

One of the great things about this particular board game is that the board can be set up differently each time you play, ensuring this is not a game you play once and then leave on a shelf in your closet. Additionally, it can increase in complexity as children grow up or become more familiar with the cards and logic of the game. For example, the "Function Frog" card is used to represent a set of several moves (i.e. it allows users to create a function or subroutine). By using this card, players learn to shorten their program by using this single card to represent a sequence of movements.

Robot Turtles can be a playful addition to family game night or used as a center activity in schools and informal education settings like camps or after-school programs. Parents and educators can take the board game concept further by encouraging children to design and create their own coding board games. This could develop into a longer STEAM curriculum unit that involves writing rules for the games, designing and decorating a playing board and/or cards, testing and improving the game by playing it with friends and peers, and more.

Books and Stories

Up to this point, we have focused on board games and hands-on materials or products to explore STEAM. But if parents and educators want to get started exploring STEAM in a very familiar way, it can be as easy as strategically integrating new books into your storytime practices. Adults can try reading and discussing engineering- or science-themed picture books and have discussions around what the characters did and why. A few notable examples include *Rosie Revere, Engineer* by Andrea Beaty; *Ada Twist, Scientist* by Andrea Beaty; *If I Built a Car* by Chris Van Dusen; *If I Built a House* by Chris Van Dusen; and *Going Places* by Paul Reynolds and Peter Reynolds.

Picture Book STEAM Activity Example

The authors developed this activity and reading list as part of their work, which included offering paid weekend-and-holiday STEM experiences for young children in the greater Boston area. Many picture books can naturally lead to a hands-on STEAM activity. For example, the book *If I Built a House* by Chris Van Dusen focuses on a child imagining the design of his dream house and all the fantastical elements it might include. After reading this book, children can create blueprints for their own dream houses, inspired by the blueprints in *If I Built a House*. Adults may also wish to facilitate a longer discussion about real-world structures, take a look at real building blueprints, and explore architecture from cultures around the world.

This can lead to a hands-on building activity that focuses on the engineering design process. The engineering design process refers to the cyclical or iterative process engineers use to design an artifact in order to meet a need. While there are many versions of the engineering design process, it typically includes a version of the following steps: identifying a problem, looking for ideas for solutions and choosing one, developing a prototype, testing, improving, and sharing solutions with others. Figure 5.5 shows one example of the engineering design process created by the Developmental Technologies Research Group at Tufts University.

After learning about the engineering design process, children can use a variety of materials such as LEGO, popsicle sticks, recycled materials, and more to bring their dream house blueprint designs to life (see Fig. 5.6). When faced with the actual materials at hand, some children may wish to revise their designs. All children should be encouraged to engage in the "test and improve" stage of the engineering design process by ensuring their houses are sturdy and implementing improvements or changes as needed.

From a STEAM perspective, this type of engineering activity can easily integrate more with fine arts, by incorporating a focus on painting, decorating, and considering the aesthetics of the houses. Or, it could integrate with literacy by connecting to a classic story such as *the Three Little Pigs*. Children could test the sturdiness of their houses against the breath of the "Big Bad Wolf" (i.e. a fan) and make any changes to their design based on the results of this test.

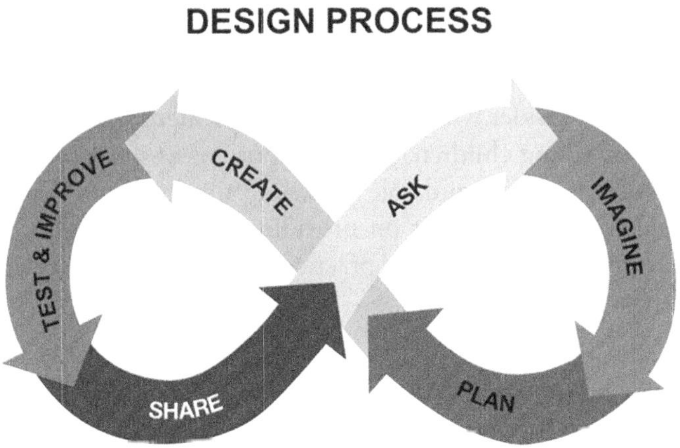

Fig. 5.5 The engineering design process

Fig. 5.6 Dream house creations made by children in K to Second Grade

Tips for Parents and Educators

Choosing appropriate tools and materials is only the beginning of what adults need to consider when it comes to implementing quality STEAM education for young children. Just as important as the tools we use, are the mindsets, attitudes, and role modeling to which we expose young children. This section focuses on providing parents and educators with tips, ideas, and resources for best practices exploring STEAM with young children.

Fostering a Growth Mindset

One of the most important things that parents and educators can do to support young children's STEAM education is fostering the right mindset toward hard work, perseverance, and failure. Research has shown that personal views about intelligence and failure may impact children's achievement and persistence in STEM fields. Psychologist Carol Dweck spent decades researching achievement and success and developed the concept of the "growth mindset" (Dweck, 2002, 2008). The "growth mindset" is the belief that intelligence is not fixed, but instead can change and grow incrementally through practice.

It is worth noting that Dweck's findings have met with criticism from the research community for her methodological approach, which her team attempted to address by launching a large-scale study of 12,000 students involving third-party research evaluators and methodological analysts. This confirmed core elements of her prior results, albeit with extremely small effects (Yeager et al., 2019). However, replication studies of Dweck's work, particularly randomized control trials, have met with mixed success. For example, Li and Bates (2017) found no achievement differences predicted by mindset in a sample of over 600 children, whereas Bettinger, Ludvigsen, Rege, Solli, and Yeager (2018) claim to have replicated Dweck's findings, and attribute their success to close adherence to Dweck's original intervention approach (see Denworth, 2019 for a full discussion of the ongoing debate about growth mindset). Despite this ongoing debate, education and psychology practitioners continue to use growth mindset in their practice, and researchers who support Dweck's work argue that educational interventions must be judged in a real-world context, where even small effects can be important (Denworth, 2019).

One way that adults can support a growth mindset is learning to praise children differently. Instead of simply telling children they are smart, which does not encourage growth, praise their *effort*. Praise the time and hard work children put into their project or mastering a new skill rather than just the outcome. Offering praise like "wow, you are so smart!" certainly can offer a short-term self-esteem boost, but in the long term, it can make children lose confidence when tasks become hard. Consider offering nuanced praise, such as, "I am so impressed that you spent so many hours working hard and building that LEGO house—I can tell it is really sturdy because of the wide base!" Not only does this type of praise help to foster the growth mindset, but it also shows you are paying close attention to their work, rather than offering generic compliments.

Role Modeling

Parents and educators should be aware of what and who young children are seeing on an everyday basis in school, at home, in the media, and in books. Are they exposed to engineers and scientists who look like them? Do they see women and minorities excelling at mathematics and using technology? Children need role models who reflect aspects of themselves that they can admire and look up to, especially within the sciences and technology. Adults can try to address this need by introducing young children to both fictional characters and real-life role models from STEAM fields that represent a range of genders and backgrounds. Some of the picture books referenced earlier in this chapter could be a great place to start. For example, *Rosie Revere, Engineer* and *Ada Twist, Scientist* both feature a female protagonist engaging in STEM.

Meeting real-world scientists and engineers can also be a powerful experience for young children. Educators can reach out to children's families for volunteers and may be surprised to find connections within your own classroom network. Parents and teachers can arrange trips to science museums, makerspaces, and laboratories for an exciting chance to meet or learn about scientists and engineers from a range of backgrounds. Local colleges and universities can also be a resource for finding diverse role models majoring in STEAM fields who may be interested in collaborating with you.

It is critical that parents and teachers do not forget about children's most impactful role models – the adults who care for them each day! Children are always watching and listening to what parents, teachers, and caregivers say and do. It is important for these adults to be modeling their own sense of scientific inquiry. How do you do this? You could start by pointing out to children when you have a hypothesis or idea that you are testing, demonstrate how you solved an engineering challenge, or share with children how mathematics or science knowledge helped you solve a problem in your everyday life. When you do not know the answer to a question a child asks, use this as an authentic opportunity to model problem-solving strategies rather than shying away from the question. In this way, you are modeling your own belief in the growth mindset and demonstrating your ability to apply the engineering and problem-solving skills you are teaching them.

Using the STEAM Resources Available

This chapter has focused on providing parents, educators, and caregivers with information about tools and approaches for teaching screen-free (and low-cost) STEAM activities to young children. But it is unlikely that anyone embarking on teaching early childhood STEAM for the first time would need to start from scratch. There are many resources available, both in-person and online. Within your own community, be sure to explore local children's libraries, museums, and makerspaces for STEAM-related resources and events. Settings like these will have access to tablets, computers, robotics kits, and other more expensive STEAM materials that you may be able to borrow or use without purchasing your own.

There are also many online resources from which parents and educators can benefit. From YouTube videos teaching the Engineering Design Process to free curriculum downloads, a variety of sites and resources support parents and early childhood educators on their STEAM journeys. For example, CS Unplugged, which was mentioned earlier in this chapter, has a range of activity and curriculum guides freely available online at csunplugged.com/. A few notable examples are as below:

- Code.org—Code.org offers many useful resources for parents and educators embarking on teaching computer science to children in grades K-12. As it relates to low-cost and screen-free activities, they have compiled a list of their unplugged curriculum ideas and resources here: code.org/curriculum/unplugged
- NASA for Educators—Lesson plans, teacher guides, classroom activities, posters and more for teachers and students as young as Kindergarten. www.nasa.gov/stem/foreducators/k-12/index.html
- Teach Engineering—A digital library comprised of standards-based engineering curricula for K-12 educators. See: www.teachengineering.org/

Conclusion

Young children are budding scientists and engineers who are naturally curious about the world around them and how things fit together and work. This means they are at a perfect age to explore STEAM and particularly, concepts of engineering and computer science. While there has been a growing focus on innovative new applications, digital devices, and software to encourage young children's exploration of computer science and engineering, there are also many low-cost and screen-free approaches to teaching the same concepts. Moreover, low-cost and low-tech materials and approaches may be useful in reaching schools and communities that are unable to afford new technologies and professional development for educators. Low-cost and low-tech STEAM approaches are accessible for parents and teachers, even those with little-to-no STEM background themselves. With this new crop of board games, card games, and unplugged activities, computer science and engineering is becoming more accessible to all.

Conflict-of-Interest Disclosure

The activities and resources presented in this chapter were developed by the authors during research-based and professional education interventions, or were available through third-party outlets (see specific examples for more information about their origins). The KIBO Robotics Kit described in the introduction is produced and marketed through KinderLab Robotics, Inc., where both authors have previously been employed. The authors are in no way affiliated with any of the other products described in the chapter.

References

Allen-Handy, A., Ifill, V., Schaar, R. Y., Rogers, M., & Woodard, M. (2020). Black girls STEAMing through dance: Inspiring STEAM literacies, STEAM identities, and positive self-concept. In *Challenges and opportunities for transforming from STEM to STEAM education* (pp. 198–219). Hershey, PA: IGI Global.

American Academy of Pediatrics. (2018). *Children and media tips from the American academy of pediatrics.*

Barnes, J., FakhrHosseini, S. M., Vasey, E., Park, C. H., & Jeon, M. (2020). Child-robot theater: Engaging elementary students in informal STEAM education using robots. *IEEE Pervasive Computing, 19*(1), 22–31.

Bell, T., Alexander, J., Freeman, I., & Grimley, M. (2009). Computer science unplugged: School students doing real computing without computers. *The New Zealand Journal of Applied Computing and Information Technology, 13*(1), 20–29.

Bell, T., & Vahrenhold, J. (2018). CS unplugged—How is it used, and does it work? In *Adventures between lower bounds and higher altitudes* (pp. 497–521). Cham, Switzerland: Springer.

Bell, T. C., Witten, I. H., & Fellows, M. (1998). *Computer Science unplugged: Off-line activities and games for all ages.* Canterbury, UK: Computer Science Unplugged.

Bers, M., Ponte, I., Juelich, K., Viera, A., & Schenker, J. (2002). Teachers as designers: Integrating robotics in early childhood education. In *Information technology in childhood education* (pp. 123–145). Morgantown, W. Va: AACE.

Bers, M. U. (2008). *Blocks to robots: Learning with technology in the early childhood classroom.* New York: Teachers College Press.

Bers, M. U. (2012). *Designing digital experiences for positive youth development: From playpen to playground.* Cary, NC: Oxford University Press.

Bers, M. U. (2018). *Coding as a playground: Programming and computational thinking in the early childhood classroom.* New York: Routledge Press.

Bers, M. U., Seddighin, S., & Sullivan, A. (2013). Ready for robotics: Bringing together the T and E of STEM in early childhood teacher education. *Journal of Technology and Teacher Education, 21*(3), 355–377.

Bettinger, E., Ludvigsen, S., Rege, M., Solli, I. F., & Yeager, D. (2018). Increasing perseverance in math: Evidence from a field experiment in Norway. *Journal of Economic Behavior & Organization, 146*, 1–15.

Brackmann, C. P., Román-González, M., Robles, G., Moreno-León, J., Casali, A., & Barone, D. (2017, November). Development of computational thinking skills through unplugged activities in primary school. In *Proceedings of the 12th workshop on primary and secondary computing education* (pp. 65–72).

Bravo Sánchez, F. Á., González Correal, A. M., & Guerrero, E. G. (2017). Interactive drama with robots for teaching non-technical subjects. *Journal of Human-Robot Interaction, 6*(2), 48–69.

Brosterman, N. (1997). *Inventing kindergarten.* New York: Henry N. Abrams.

Busteed, B., & Sorenson, S. (2015, August 20). *Many students lack access to computer science learning*. Retrieved from: www.gallup.com/education/243416/students-lack-access-computer-science-learning.aspx

Cejka, E., Rogers, C., & Portsmore, M. (2006). Kindergarten robotics: Using robotics to motivate math, science, and engineering literacy in elementary school. *International Journal of Engineering Education, 22*(4), 711.

Clements, D. H. (1999). The future of educational computing research: The case of computer programming. *Information Technology in Childhood Education Annual, 1*, 147–179.

Corbett, C., & Hill, C. (2015). *Solving the equation: The variables for women's success in engineering and computing*. Washington, DC: American Association of University Women.

Curzon, P., McOwan, P. W., Plant, N., & Meagher, L. R. (2014, November). Introducing teachers to computational thinking using unplugged storytelling. In *Proceedings of the 9th workshop in primary and secondary computing education* (pp. 89–92). Association for Computing Machinery (ACM).

Denworth, J. (2019). Debate arises over teaching "growth mindsets" to students. *Scientific American*. Retrieved from https://www.scientificamerican.com/article/debate-arises-overteaching-growth-mindsets-to-motivate-students

Dweck, C. S. (2002). The development of ability conceptions. In *Development of achievement motivation* (pp. 57–88). Cambridge, MA: Academic Press.

Dweck, C. S. (2008). *Mindset: The new psychology of success*. New York: Random House Digital Inc.

Elkin, M., Sullivan, A., & Bers, M. U. (2016). Programming with the KIBO robotics kit in preschool classrooms. *Computers in the Schools, 33*(3), 169–186. https://doi.org/10.1080/07380569.2016.1216251

Fessakis, G., Gouli, E., & Mavroudi, E. (2013). Problem solving by 5–6 years old kindergarten children in a computer programming environment: A case study. *Computers & Education, 63*, 87–97.

Greiff, S., Wüstenberg, S., Csapó, B., Demetriou, A., Hautamäki, J., Graesser, A. C., et al. (2014). Domain-general problem solving skills and education in the 21st century. *Educational Research Review, 13*, 74–83.

Hickmott, D., Prieto-Rodriguez, E., & Holmes, K. (2018). A scoping review of studies on computational thinking in K–12 mathematics classrooms. *Digital Experiences in Mathematics Education, 4*(1), 48–69.

Kazakoff, E., Sullivan, A., & Bers, M. U. (2013). The effect of a classroom-based intensive robotics and programming workshop on sequencing ability in early childhood. *Early Childhood Education Journal, 41*(4), 245–255. https://doi.org/10.1007/s10643-012-0554-5

Kim, J. O., & Kim, J. (2020). Development and application of art based STEAM education program using educational robot. In *Robotic systems: Concepts, methodologies, tools, and applications* (pp. 1675–1687). Hershey, PA: IGI Global.

Kuh, L. P. (Ed.). (2014). *Thinking critically about environments for young children: Bridging theory and practice*. New York: Teachers College Press.

Lee, K., Sullivan, A., & Bers, M. U. (2013). Collaboration by design: Using robotics to foster social interaction in kindergarten. *Computers in the Schools, 30*(3), 271–281.

Li, Y., & Bates, T. (2017). *Does growth mindset improve children's IQ, educational attainment or response to setbacks? Active-control interventions and data on children's own mindsets*. Edinburgh, UK: University of Edinburgh. https://osf.io/preprints/socarxiv/tsdwy/

Madill, H., Campbell, R. G., Cullen, D. M., Armour, M. A., Einsiedel, A. A., Ciccocioppo, A. L., et al. (2007). Developing career commitment in STEM-related fields: Myth versus reality. In R. J. Burke, M. C. Mattis, & E. Elgar (Eds.), *Women and minorities in science, technology, engineering and mathematics: Upping the numbers* (pp. 210–244). Northampton, MA: Edward Elgar Publishing.

Markert, L. R. (1996). Gender related to success in science and technology. *The Journal of Technology Studies, 22*(2), 21–29.

McClure, E. R., Guernsey, L., Clements, D. H., Bales, S. N., Nichols, J., Kendall-Taylor, N., et al. (2017). STEM starts early: Grounding science, technology, engineering, and math education in early childhood. In *Joan Ganz Cooney Center at Sesame Workshop*. New York: Joan Ganz Cooney Center at Sesame Workshop.

Messer, D., Thomas, L., Holliman, A., & Kucirkova, N. (2018). Evaluating the effectiveness of an educational programming intervention on children's mathematics skills, spatial awareness and working memory. *Education and Information Technologies, 23*(6), 2879–2888.

Metz, S. S. (2007). Attracting the engineering of 2020 today. In R. J. Burke, M. C. Mattis, & E. Elgar (Eds.), *Women and minorities in science, technology, engineering and mathematics: Upping the numbers* (pp. 184–209). Northampton, MA: Edward Elgar Publishing.

Moomaw, S., & Davis, J. A. (2010). STEM comes to preschool. *YC Young Children, 65*(5), 12.

National Science and Technology Council. (2018). *Charting a course for success: America's strategy for STEM education*. Washington, DC. Retrieved from: www.whitehouse.gov/wp-content/uploads/2018/12/STEM-Education-Strategic-Plan-2018.pdf

Nicholson, S. (1972). The theory of loose parts, an important principle for design methodology. *Studies in Design Education Craft & Technology, 4*(2), 5–14.

Perlman, R. (1976). *Using computer technology to provide a creative learning environment for preschool children*. Cambridge, MA: MIT, A.I. Laboratory.

Resnick, M., Martin, F., Berg, R., Borovoy, R., Colella, V., Kramer, K., & Silverman, B. (1998, January). Digital manipulatives: New toys to think with. In *Proceedings of the SIGCHI conference on human factors in computing systems* (pp. 281–287). United States: ACM Press/Addison-Wesley Publishing Co.

Rodriguez, B., Rader, C., & Camp, T. (2016, July). Using student performance to assess CS unplugged activities in a classroom environment. In *proceedings of the 2016 ACM conference on innovation and technology in computer science education* (pp. 95–100). New York, NY, United States: Association for Computing Machinery

Scherer, R. (2016). Learning from the past–the need for empirical evidence on the transfer effects of computer programming skills. *Frontiers in Psychology, 7*, 1390.

Scherer, R., Siddiq, F., & Sánchez Viveros, B. (2019). The cognitive benefits of learning computer programming: A meta-analysis of transfer effects. *Journal of Educational Psychology, 111*(5), 764.

Sentance, S., & Csizmadia, A. (2017). Computing in the curriculum: Challenges and strategies from a teacher's perspective. *Education and Information Technologies, 22*(2), 469–495.

Shapiro, D. (2014, August). *Robot turtles: The board game for little programmers*. www.kickstarter.com/projects/danshapiro/robot-turtles-the-board-game-for-little-programmer/posts/597883

Smith, N., Allsop, Y., Caldwell, H., Hill, D., Dimitriadi, Y., & Csizmadia, A. P. (2015, November). Master teachers in computing: What have we achieved?. In *Proceedings of the Workshop in Primary and Secondary Computing Education* (pp. 21–24).

Steele, C. M. (1997). A threat in the air: How stereotypes shape intellectual identity and performance. *American Psychologist, 52*, 613–629.

Strawhacker, A. L., Lee, M. S. C., & Bers, M. U. (2017). Teaching tools, teachers' rules: Exploring the impact of teaching styles on young children's pro-

gramming knowledge in ScratchJr. *International Journal of Technology and Design Education*. https://doi.org/10.1007/s10798-017-9400-9
Sullivan, A., & Bers, M. U. (2015). Robotics in the early childhood classroom: Learning outcomes from an 8-week robotics curriculum in pre-kindergarten through second grade. *International Journal of Technology and Design Education*.
Sullivan, A., & Bers, M. U. (2017). Dancing robots: Integrating art, music, and robotics in Singapore's early childhood centers. *International Journal of Technology and Design Education*. Advance online publication. doi:https://doi.org/10.1007/s10798-017-9397-0
Sullivan, A., Bers, M. U., Mihm, C. (2017). Imagining, playing, & coding with KIBO: Using KIBO robotics to Foster computational thinking in young children. In *Proceedings of the international conference on computational thinking education*. Wanchai, Hong Kong.
Sullivan, A. A. (2019). *Breaking the STEM stereotype: Reaching girls in early childhood*. Lanham, MD: Rowman & Littlefield Publishers.
U.K. Department for Education. (September, 2013). *National curriculum in England: Computing programmes of study. Statutory guidance*. London: Crown copyright.
U.S. Department of Education. (2010). *Transforming American education: Learning powered by technology*. Washington, DC. Retrieved from: www.ed.gov/technology/netp-2010
Watson, E. (2020). STEM or STEAM?: The critical role of arts in technology education (and the critical role of art in technology). *Irish Journal of Academic Practice, 8*(1), 8.
Wyeth, P. (2008). How young children learn to program with sensor, action, and logic blocks. *The Journal of the Learning Sciences, 17*(4), 517–550.
Yakman, G. (2008, February). STEAM education: An overview of creating a model of integrative education. In *Pupils' attitudes towards technology (PATT-19) conference: Research on technology*. Salt Lake City, Utah: Innovation, Design & Engineering Teaching.
Yeager, D. S., Hanselman, P., Walton, G. M., Murray, J. S., Crosnoe, R., Muller, C., et al. (2019). A national experiment reveals where a growth mindset improves achievement. *Nature, 573*(7774), 364–369.
Zviel-Girshin, R., Luria, A., & Shaham, C. (2020). Robotics as a tool to enhance technological thinking in early childhood. *Journal of Science Education and Technology, 29*, 294–302.

6

Integrating Engineering Within Early STEM and STEAM Education

Lyn D. English

Why Early Engineering?

Engineering learning in kindergarten and the beginning school years is an underrepresented field, yet it is one of the most practical and real-world learning domains that all children can experience. Indeed, many of the activities that children naturally engage in at play both at school and at home are rich examples of early engineering. Because engineering shapes so much of our actual and virtual worlds, it is an ideal discipline to both link and promote the varied capabilities that young children bring to informal and formal learning environments.

Early engineering draws on both the content and processes of engineering, with the foundational content concerned with aspects of our human-made world (Evangelou & Bagiati, 2019). As such, early

L. D. English (✉)
Queensland University of Technology, Brisbane, QLD, Australia
e-mail: l.english@qut.edu.au

C. Cohrssen, S. Garvis (eds.), *Embedding STEAM in Early Childhood Education and Care*, https://doi.org/10.1007/978-3-030-65624-9_6

engineering offers a range of topics suitable for young learners, topics that span multiple disciplines, not just mathematics, science, and technology. Engineering-based activities can thus be developed to fit in with a school's curriculum, community projects, children's home environments, and local playgrounds.

If you've ever watched children at play, then you know they're fascinated with building things—and with taking things apart to see how they work. In other words, children are natural-born engineers (Cunningham et al., 2020).

Given the relevance and adaptability of engineering to early learning, one wonders why the E in STEM remains mostly silent, especially during the kindergarten and early school years. As Cunningham emphasised several years ago,[1] children learn about the natural world while studying science, "but what about the human-made world built on top of it—the buildings and vehicles and screens where they spend the vast majority of their time?" Much of the world in which we live is designed by engineers, so it is of considerable concern that the discipline is largely ignored until late secondary or university levels in many nations. Several possible reasons exist for this, including a belief that children do not have the capabilities to engage with such an "advanced" discipline knowledge, an apparent lack of resources for implementing early engineering, and limited teacher development programmes. While the availability of resources has increased in recent years, especially in the United States, teacher preparation remains problematic, with many teachers thus lacking confidence in engineering and STEM education more broadly (Crismond & Adams, 2012; National Academies of Sciences, Engineering, and Medicine, 2020). Of particular concern to this chapter, however, is the underestimation of children's capabilities for the discipline—a view that is a potential barrier to implementing engineering.

Developmentally appropriate engineering experiences remain a missing link in children's early education (Lippard, Lamm, Tank, & Choi, 2019). To ignore the STEM capabilities of young learners is to deny them the rich learning of which they are capable. Unfortunately, in

[1] https://www.discovermagazine.com/the-sciences/teaching-kids-to-think-like-engineers#.UwZzZvldXh4.

underestimating what young learners can achieve, educators tend to remain with basic activities that fail to adequately meet learners' potential (Engel, Claessens, Watts, & Farkas, 2016). The learning capabilities of children thus remain untapped, with few opportunities to extend their vibrant kindergarten and school-entry knowledge (MacDonald & Carmichael, 2018).

Young learners are eminently capable of working with engineering-based problems—their natural curiosity, inquiry, and desire to explore their world form not only the cornerstone of early childhood development (Brophy, Klein, Portsmore, & Rogers, 2008), but also are a key element of "thinking like an engineer" (Elkin, Sullivan, & Bers, 2018). To ignore this early childhood potential is to deny children the diverse experiences that engineering can provide. The urgent need for more developmentally and culturally appropriate curriculum resources in engineering remains a pressing concern (Early Childhood STEM Working Group, 2017). We need to increase awareness of, and capitalise on, children's skills as independent problem-solvers, who relish challenges, persevere in the face of failure, and learn from both what "works" and what does not. Educators, including parents, should be cognizant of how children's talents can be harnessed and enriched to sow the seeds of engineering education.

Early childhood curricula seldom highlight the ease with which engineering can be integrated with the other STEM disciplines, as well as with the arts and aesthetics more broadly, including literature. As Petroski (2016) highlighted, "Engineering is not an end in itself. It operates in a moral, social, economic, and aesthetic context" (p. 21). Engineering is thus an ideal discipline for developing STEAM programmes, with its focus on the design and creation of human-made products and processes within the contexts that Petroski describes (National Academies of Sciences, Engineering, and Medicine, 2020).

Early Engineering Literacy

A recent report from the Academies of Sciences, Engineering, and Medicine (2020) defined an engineering literate person as one who "has a basic understanding of the people and processes involved in creating the human-built world" (p. 47). With this foundation, one can think critically and make decisions about a range of issues important to one's life, family, and community. Drawing on general principles for K-12 engineering education (NAE & NRC, 2009), the Academies highlighted a number of core features of engineering literacy. These include developmentally appropriate mathematics, science, and technology skills, and the promotion of "engineering habits of mind"—ways of thinking and acting that are important for both early engineering learning and overall school readiness (Lippard et al., 2019). Additional features can be added to engineering literacy for children, namely, an appreciation of the work of engineers and engineering in the immediate and wider world, a curiosity for learning "how things work", an awareness of how engineering draws on the other STEM disciplines, and a keenness to solve hands-on, real-world problems for which multiple solutions are possible (English & Moore, 2018). Implicit in these components is an appreciation of aesthetics in engineering, an important feature to which I return.

Engineering Design

Engineering design provides foundational, interdisciplinary links that are invaluable in developing modelling and problem-solving capabilities across the STEM fields (English, 2018a, b). Engineering design is often referred to as the "interdisciplinary glue" that facilitates learning in the other STEM disciplines, as well as in other domains such as literature and engineering (Tank et al., 2018, p. 175). Yet the multiple applications of engineering design in the curriculum are not being acknowledged adequately, let alone utilised.

Numerous descriptions of engineering design have appeared in the literature (McCormick & Hammer, 2016; Watkins, Spencer, & Hammer, 2014). Engineering design is usually defined as iterative in nature and comprising processes such as identifying problem goals, brainstorming ideas, meeting problem constraints, sketching designs, balancing trade-offs, predicting possible outcomes, testing and assessing initial constructions, and redesigning and reconstructing. Although engineering design encompasses complex, multi-faceted processes, it is readily accessible to young learners (Purzer & Douglas, 2018). For example, Tank et al. (2018) developed a representation of engineering design for the beginning school grades that consisted of six main components divided into problem scoping (define and learn) and solution generation (plan, try out, test, and decide [whether a solution meets the criteria and constraints, or whether a redesign is needed]). Each design component is linked with the important processes of communication and teamwork. These latter processes form part of the engineering habits of mind, which go hand-in-hand with design processes.

In another recent study, Tank, Rynearson and Moore (2018) investigated how an engineering design-based STEM integration unit was enacted across three kindergarten classrooms. Their results suggested that kindergarten children were able to engage meaningfully in, and with, multiple phases of engineering design while also developing an understanding of the work of scientists and engineers. Through multiple instances of child-initiated talk, interactions between children, and the use of explicit engineering language, the children were observed to make connections to prior learning. The research further revealed that engineering design activities in the early school years should incorporate multiple aspects of engineering and engineering design, together with interdisciplinary content and a context for STEM integration. Tank, Moore, et al. (2018) concluded that kindergarten children are able to complete long-term, multi-component engineering design projects that incorporate integrated STEM lessons, and that young children can develop high levels of understanding and engagement. The researchers warned, however, that teachers need to facilitate and guide their students in their learning during an engineering design-based project, but at the same time, not constrain them. Such an approach calls for preservice and in-service teacher programmes that guide teachers in the implementation of activities incorporating engineering design processes.

Engineering Habits of Mind

Engineering habits of mind encompass several of the so-called twenty-first century skills (P21, 2015) and include *systems thinking, creativity, optimism, collaboration, communication*, and attention to *ethical considerations* (Katehi, Pearson, & Feder, 2009; Lippard, Riley, & Lamm, 2018; National Academy of Engineering & National Research Council, 2009; National Academies of Sciences, Engineering, and Medicine, 2020, p. 30). Like design processes, engineering habits of mind (Lippard et al., 2019) are powerful thinking skills that are applicable across all of the STEM disciplines as well as other domains. Systems thinking in particular, as a twenty-first century skill, is a significant habit of mind that can be fostered as children recognise disciplinary links within their engineering experiences. Children learn how interactions among the components of an engineering problem (the system) can have unanticipated effects on the solution (e.g., how a child places individual building blocks in creating a chosen structure can lead to a stable or unstable product). Children are capable of *systems thinking* (Lippard et al., 2018), yet this skill is underrepresented in our education programmes (Salado, Chowdhury, & Norton, 2018). Indeed, systems thinking is essential to dealing with world problems that impact us all, such as global economies, the spread of viruses, population growth, and overcrowding of cities. Systems thinking should be introduced early through the provision of challenging problems that involve interactions among parts impacting on the whole. Yet as Salado et al. (2018) indicate, we know little about the developmental or cognitive components of systems thinking.

Engineering naturally engenders *creativity* in children, as it fosters the use of "imagination in the design processes of engineering" (Loveland & Dunn, 2014, p. 14). As evidenced in the Palmer and van Bommel (2018) study, young children are both creative and competent as they engage in challenging mathematical and engineering-based tasks for which they do not seem to need any kind of "special preparation" (p. 1788). Engineering-based activities promote creativity as children consider, for example, the limitations and possibilities of construction materials available to them

(Lippard et al., 2018), as they design and plan possible solutions, and as they create a desired product.

Engineering experiences foster *optimism*, empowering all children to experience a sense of success in generating their own solution to a problem for which multiple solutions are possible. The iterative processes of design enable children to create and test prototypes to determine if problem requirements and constraints are met. Opportunities to redesign, if necessary, provide a basis for optimism to flourish as children persevere in their efforts to improve their creations. Encouraging optimistic dispositions is critical in children's learning more broadly, where possessing the motivation to engage and remain engaged in challenging tasks and concepts should be developed early. Indeed, optimism is frequently observed in young children's general play activities, yet such an important habit of mind can frequently fall by the wayside as children progress to the higher year levels (Potvin & Hasni, 2014).

Both *collaboration* and *communication* are essential for children to participate productively in group engineering activities as well as in whole-class discussions about their creations. Effective communication is a primary requirement of productive group collaboration. Through interacting with group members, children are afforded opportunities to strengthen their STEM knowledge, appreciate the different viewpoints of group members, and share, debate and refine individual and group ideas (English, 2017; Loveland & Dunn, 2014). Explaining ideas, justifying and defending arguments, and developing an in-depth understanding of different solution approaches are foundational to early engineering learning as well as to STEM more broadly.

Communication within and between groups is not the only way children communicate their ideas in early engineering—their design sketches also play a vital role in communication. Design sketches help develop and convey meaning and understanding about a problem (Anning, 1997; English, 2017; Song & Agogino, 2004) to their peers. Young children's simple sketches communicate how their desired artefact will be constructed; the inclusion of a range of annotations, such as measurements, orientations, materials use, and construction steps, enhances communication.

The final engineering habit of mind pertains to *ethical considerations*, where "any given solution to a problem will impact others in the environment and the environment itself" (Lippard et al., 2018, p. 26). This habit of mind is a sophisticated one for young children and usually requires adult scaffolding, such as encouraging children to make predictions and anticipate possible outcomes or impacts of their engineering actions (Lippard et al., 2018). Consideration of ethical factors can be further heightened when children discuss, for example, new projects or constructions in their environment that could have a potential negative impact on people's lives.

In sum, by incorporating engineering design and engineering habits of mind within an early learning curriculum, young children's conceptual understanding across the STEM disciplines can be fostered (English, 2018b). Ultimately, such conceptual learning should lead to the development of what McKenna (2014) refers to as "adaptive expertise" (p. 232) where learning from one problem-solving activity is adapted and applied to new situations. In other words, engineering-based problems have the potential to encourage young students to "learn from and about the problem, while continually reflecting on, and possibly reshaping, prior knowledge and experiences" (McKenna, 2014, p. 232). Early engineering-based problems that embed design constraints and draw on meaningful interdisciplinary contexts can assist learners to recognise what knowledge they need to apply to a new problem situation.

The Arts and Engineering

The importance of the arts (including literature) in engineering learning has not been realised as fully as it should. While some use has been made of literature, which provides a natural companion to STEM learning, the arts and aesthetic experiences more broadly remain under-utilised and under-researched (English, 2018b; Ostrowski, Becker, Diaz Caceres, Lam-Herrera, & Rothschuh, 2018; Sinclair, 2006). Integration of the arts within STEM is readily accomplished when addressing early engineering. By its very nature, early engineering is aesthetically enticing, as are numerous engineering feats across the globe. Indeed, we can find endless examples of engineering creations in our world that are designed to be both functional and aesthetically appealing. Jorn Utzon, for example, designed the Sydney Opera House (Australia), for which Sydney is famous. He was inspired by nature, its forms, its functions, and colours in creating his design which was influenced by bird wings and the shape and form of clouds, shells, palm trees, and walnuts. In drawing on nature, Utzon designed a structure that was functional, sustainable, efficient, and aesthetically pleasing.

Research on aesthetics in children's mathematical learning (e.g., Sinclair, 2006) has revealed how young learners can find beauty in the mathematics they experience, which can override a preoccupation with "correctness" and "passing tests". Fortunately, early engineering provides opportunities for aesthetically pleasing experiences where learners can create artefacts that hold beauty and personal meaning for them. On the other hand, adding another discipline to the STEM acronym may promote further compartmentalisation of its disciplines, rather than more cohesiveness in their integration (Ostrowski et al., 2018). At the same time, integrating five disciplines presents the danger of watering down each of them. Clearly, further research is needed on this issue. This debate is further addressed in this book.

One component of the arts that integrates effectively with engineering is literature. The appealing illustrations and stories in children's picture books provide powerful contexts for stimulating engineering experiences—whether it be general children's literature (non-engineering specific) or engineering-specific literature (English, 2018b). Both types provide valuable resources, with the latter becoming more prevalent in recent years (e.g., King & Johnston, 2014). Research has indicated that combining the engineering-centred literature with classroom discussions can be a powerful way to broaden children's participation in engineering education (Cunningham, Lachapelle, & Davis, 2018; Pantoya, Aguirre-Munoz, & Hunt, 2015). Pantoya et al., for example, developed an engineering-centred literacy programme that can be integrated within, or complement, national science curricula. Implementing the programme across pre-K to second grade, the authors incorporated a concept sketching activity to develop an engineering identity, that is, to help the children learn about the work of different engineers who shape our world. The storybook, *Engineering Elephants* (Hunt & Pantoya, 2010), engaged children in an interactive journey of an elephant as he questioned the world around him. The study findings show how their engineering-centred literature not only enhanced the children's creative skills but also developed an appreciation of the nature and role of engineers and their work. In doing so, the programme prompted an early interest in, and enthusiasm for, engineering-based activities.

Other engineering-centred children's literature includes the series developed by King and Johnston (e.g., 2014), which takes children on a journey with human and animal characters as they solve appealing problems. Children are encouraged to contribute to problem solution as each new scenario is presented, with hypotheses considered. The problem scenarios in this series are primarily engineering-based but also incorporate the other STEM areas. The series is aesthetically pleasing with the artwork not only designed to captivate young children's interest but also to support the problems presented and provide options for possible solutions. Figure 6.1 provides an example of the way in which the artwork serves these two primary purposes.

Fig. 6.1 The front cover of *Engibear's Bridge* (2014), with kind permission from King and Johnston

Examples of Engineering Within STEAM for Kindergarten and the Home

Block building is a common activity both in kindergarten and in the home. Observations of children's free block play in kindergarten classrooms have shown that their actions reflect similarities to engineering design processes, as well as a range of engineering habits of mind. Bagiati and Evangelou's (2016) qualitative observational study of 18 preschoolers' free play with blocks over a period of four months demonstrated instances of early engineering behaviour. Videotaped data showed the children articulating goal-oriented behaviour, problem-solving processes, innovative actions, multiple approaches to designing, pattern repetition, and testing and improving.

Given that building blocks are a staple of early childhood centres and are frequently found in children's home, they are an ideal tool for fostering early engineering. In particular, building blocks can serve as an informal introduction to engineering construction and design processes. Within a home setting, parents/carers can observe children's block play to identify their early engagement with engineering processes. For example, children often identify a "pretend" problem they are trying to solve (e.g., building a house, a bridge, a tall tower, etc.), set goals they wish to achieve, share their ideas, create their desired product, test their constructions, and improve on their initial creations (Bagiati & Evangelou, 2016). Parents can question their children with queries such as the following: *Tell me what you have built. What does this part of your* (construction) *do? Why did you place that block/s there? What might have happened if you had placed these blocks over here? How did you stop your* (construction) *from toppling over? How did you make it balance? Was your (construction) easy or difficult to build? Why? Did you need to place some blocks in a different position?* And *what might you do to make your* (construction) *even better?*

A more advanced block activity appears in *Family Engineering: An Activity & Event Planning Guide* (2011). Using 14 tubes (e.g., empty toilet paper rolls) and 3 squares of corrugated cardboard (each 30 cm x 30 cm), children are to build a tower as displayed in Fig. 6.2. Given the following constraints, children are required to take turns in removing one

Fig. 6.2 Problem activity involving tower building

tube at a time from the tower, without letting the three cardboard platforms fall: (a) both hands may be used; (b) the cardboard platforms may only be touched when removing or moving a tube; and (c) the position of the remaining tubes may be changed. Following the activity, children describe the steps they took to keep the tower from tumbling. When/if their tower eventually falls, then children can explain what caused it to fall.

Other productive engineering experiences for both kindergarten and the home include the use of children's literature to foster engineering and STEM learning more broadly. Such interdisciplinary experiences promote both engineering and literacy development, as well as learning in other STEM areas. In the *Novel Engineering* programme (Portsmore & Milto, 2018), for example, children's literature is used as the basis and stimulus for engineering design challenges, drawing details from the book to identify the characters facing a problem and the constraints they confront in finding a solution. With multiple interpretations, a book's text can foster interactive idea generation where children connect their own experiences in helping the central character/s solve the dilemma being faced.

Typically, a teacher (or parent) would stop the reading of the book at appropriate points, giving children opportunities to identify the character/s' problem, the actions or situations that led to the problem, and to define the challenge being met. Applying design processes, children can discuss what is needed in solving the dilemma, steps that might be taken, and simple materials that might be needed. Sketching a design of a possible solution and using the sketch to help solve the dilemma would follow. For example, the new storybook, *The Goat Café* (Simon & Broadley, 2019), relates numerous problems that arose when a garden gate was left open and goats escaped, heading towards a small, country café to begin their adventure. Given that goats love to eat anything and everything, the goats not only devoured all the fresh food in the café, but also the café signs, the garden bed, a shed, the surrounding shrubs, and neighbouring fences. The story provides an appealing context for children to identify numerous problems that arose, and might arise, for the characters involved. Children can suggest possible approaches to resolving the problems and subsequently construct their solutions (e.g., using building materials to design and construct a new "goat-proof" fence, or design a new "goat café" with new signs and menus).

Suggested Resources

There is an increasing range of resources to help teachers and parents in promoting engineering learning in early childhood. Several worthwhile online resources include *eGFI* (http://teachers.egfi-k12.org/), *TeachEngineering* (https://www.teachengineering.org/), *Engineering is*

Elementary (*EiE*; https://www.eie.org/), and *PictureSTEM* (http://picturestem.org/picturestem-units/). These sites contain rich examples of engineering activities including those designed for kindergarten children. For example, the *TeachEngineering* website incorporates downloadable activities for homeschooling, while *PictureSTEM* comprises lesson plans for STEM activities across K-2 that focus on the engineering design process.

The programme, *Engineering is Elementary*, incorporates "EIE for Kindergarten" (https://info.eie.org/eie-k) as well as "Wee Engineer" (https://info.eie.org/wee-engineer), both of which cater to very young children. The main programme, *Engineering is Elementary*, is the original programme and targets grades K-5. *Wee Engineer* the latest component comprises activities that guide children in applying a simplified version of the engineering design process: Explore (Find out more), Create (try an idea), and Improve (make it better). *EiE for Kindergarten* focuses on strong foundations for problem-solving and critical thinking designed to prepare children for success in school and life.

A delightful new book kit, *Young Engineers*, has been created by King, Lewin, and Johnston (2019). The kit comprises the picture storybook, *Young Engineers*, a CD of songs about engineering for young children, and an activity book, *Young Engineers Projects: Dream It, STEAM It*. The activity book contains a collection of rich, easily accessible engineering activities, which cover a range of engineering fields that are directly relevant to the world of young children. The activities include background information on the particular engineering field/s, a description of the activity itself, and preparation, procedures, related experiences, and helpful hints on activity set-up.

Future Developments

Although the introduction of early engineering within kindergarten and the early school years has yet to gain substantial momentum in many countries, there is growing interest in the field especially in the opportunities it provides for linking the STEM disciplines (e.g., English, 2018a, b; Jurgenson & Delaney, 2020). Further advancement is needed, however.

It is imperative that we begin with a greater awareness of young children's competencies for undertaking early engineering activities. Although research has repeatedly shown what young learners can achieve when challenged with new experiences, their talents often remain unseen, especially when school-entry measures or other such benchmarks are used to determine what they can accomplish. Research that focuses solely on children's current knowledge and reasoning, rather than the extent of their capability is inadequate (English, 2016; Ginsburg, 2016). Awareness of young learners' potential, needs to be accompanied by more professional development and an increase in teacher resources for developing early engineering.

There is more to professional development, however, than just providing one-off sessions to introduce a new programme. Estapa and Tank (2017) report research that highlights the need for sustained, coherent, and collaborative teacher programmes, which develop more in-depth understandings of the STEM domains and various ways of integrating engineering within regular programmes. Assisting teachers to better understand the nature and role of engineering learning, together with effective planning and enactment of integrated STEM lessons, appears a key area for future research and action. At the same time, changes in policy that inform curriculum improvement are required.

References

Anning, A. (1997). Drawing out ideas: Graphicacy and young children. *International Journal of Technology and Design Education, 7*, 219–239.

Bagiati, A., & Evangelou, D. (2016). Engineering Curriculum in the Preschool Classroom: The Teacher's Experience. European Early Childhood Education Research Journal, 23, 112–128. https://doi.org/10.1080/1350293X.2014.991099.

Brophy, S., Klein, S., Portsmore, M., & Rogers, C. (2008). Advancing engineering education in P-12 classrooms. *Journal of Engineering Education, 97*(3), 369–387.

Crismond, D. P., & Adams, R. S. (2012). The informed design teaching and learning matrix. *Journal of Engineering Education, 101*(4), 738–797.

Cunningham, C. M., Lachapelle, & Davis, M. E. (2018). Engineering concepts, practices, and trajectories for early childhood education. In L. D. English & T. J. Moore (Eds.), *Early engineering learning* (pp. 273–284). Singapore: Springer.

Cunningham, C. M., Lachapelle, C. P., Brennan, R. T., Kelly, G. J., San Antonio Tunis, C., & Gentry, C. A. (2020). The impact of engineering curriculum design principles on elementary students' engineering and science learning. *Journal of Research in Science Teaching*. 57, 423–453 https://doi.org/10.1002/tea.21601.

Early Childhood STEM Working Group. (2017). *Early STEM matters: Providing high-quality STEM experiences for all young learners.* http://ecstem.uchicago.edu

Elkin, M., Sullivan, A., & Bers, M. (2018). Books, butterflies, and 'bots: Integrating engineering and robotics into early childhood curricula. In L. D. English & T. J. Moore (Eds.), *Early engineering learning* (pp. 225–248). Singapore: Springer.

Engel, M., Claessens, A., Watts, T., & Farkas, G. (2016). Mathematics content coverage and student learning in kindergarten. *Educational Researcher, 45*(5), 293–300.

English, L. D. (2016). Developing Early Foundations through Modeling with Data. In C. Hirsch (Ed.). *Annual perspectives in mathematics education: Mathematical modeling and modeling mathematics* (pp. 187–195). Reston, VA: National Council of Teachers of Mathematics.

English, L. D. (2017). Advancing elementary and middle school STEM education. *International Journal of Science and Mathematics Education*, 15, Supplement 1, 5–24.

English, L. D. (2018a). Learning while designing in a fourth-grade integrated STEM problem. *International Journal of Technology and Design Education.* https://doi.org/10.1007/s10798-018-9482-z

English, L. D. (2018b). Disruption and learning innovation across STEM. Plenary presented at the *5th International Conference of STEM in Education*. Brisbane. https://stem-in-ed2018.com.au/proceedings-2/.

English, L. D., & Moore T. (Eds.), (2018). Early Engineering Learning. Singapore: Springer. https://doi.org/10.1007/978-981-10-8621-2_4.

Estapa, A. T., & Tank, K. M. (2017). Supporting integrated STEM in the elementary classroom: A professional development approach centered on an engineering design challenge. *International Journal of STEM Education, 4*(6). https://doi.org/10.1186/s40594-017-0058-3

Evangelou, D., & Bagiati, A. (2019). Engineering in early learning environments. In L. Cohen & S. Whyte-Stupianski (Eds.), *STEM in early childhood education: How science, technology, engineering, and mathematics strengthen learning* (pp. 46–62). Abingdon, UK: Taylor & Frances.

Family Engineering. (2011). *Foundation for family science & engineering and Michigan Technological University.*

Ginsburg, H. P. (2016). Helping early childhood educators to understand and assess young children's mathematical minds. *ZDM Mathematics Education, 48*(7), 491.

Hunt, E. M., & Pantoya, M.L. (2010). *Engineering Elephants.* Author House Publishing, ISBN: 978-1-4490-5816-6.

Jurgenson, K. N., & Delaney, A. R. (2020). The three little engineers. *Mathematics Teacher: Learning & Teaching, PK-12, 113*(2), 110–116.

Katehi, L., Pearson, G., & Feder, M. (2009). The status and nature of K-12 engineering education in the United States. *The Bridge: Linking Engineering and Society, 39*(3), 5–10. https://www.nae.edu/Publications/Bridge/16145/16161.aspx

King, A., & Johnston, B. (illustrator). (2014). *Engibear's bridge.* Frenchs Forest, NSW: Little Steps Publishing.

King, A., Lewin, S., & Johnston, B. (2019). *Young Engineers (and young engineers songs, and young engineers projects: Dream it, steam it).* Frenchs Forest, Australia: Little Steps Publishing.

Lippard, C. N., Lamm, M. H., Tank, C. M., & Choi, J. Y. (2019). Pre-engineering thinking and enginering habits of mind in pre-school classroom. *Early Childhood Education Journal, 47*, 187–198.

Lippard, C. N., Riley, K. L., & Lamm, M. H. (2018). Encouraging the development of engineering habits of mind in prekindergarten learners. In L. D. English & T. Moore (Eds.), *Early engineering learning* (pp. 19–36). Singapore: Springer.

Loveland, T. & Dunn, D. (2014). Technology and Engineering Teacher, 73 (8), 13–19.

MacDonald, A., & Carmichael, C. (2018). Early mathematical competencies and later achievement: Insights from the longitudinal study of Australian children. *Mathematics Education Research Journal, 30*(4), 429–444.

McCormick, M., & Hammer, D. (2016). Stable beginnings in engineering design. *Journal of Pre-College Engineering Education Research, 6*(1), 4.

McKenna, A. F. (2014). Adaptive expertise and knowledge fluency in design and innovation. In A. Johri & B. M. Olds (Eds.), *Cambridge handbook of engi-*

neering education research (pp. 227–242). New York: Cambridge University Press.

National Academies of Sciences, Engineering, and Medicine. (2020). *Building capacity for teaching engineering in K-12 education*. Washington, DC: The National Academies Press. https://doi.org/10.17226/25612

National Academy of Engineering, & National Research Council. (2009). Engineering in K-12 education: Understanding the status and improving the prospects. Washington, DC: The National Academies Press. https://doi.org/10.17226/12635.

Ostrowski, C. P., Becker, S., Diaz Caceres, Z., Lam-Herrera, M., & Rothschuh, S. (2018). Exploring the margins of the field: Rethinking STEM in education. In *Proceedings of the 13th international conference of the learning Sciences* (pp. 1105–1108). https://doi.dx.org/10.22318/cscl2018.1105

P21 (2015). *P21 framework for 21st century learning*. The partnership for 21st century learning.

Palmer, H., & van Bommel, J. (2018). Problem solving in early mathematics teaching—A way to promote creativity? *Creative Education, 9*, 1775–1793.

Pantoya, M. L., Aguirre-Munoz, Z., & Hunt, E. M. (2015). Developing an engineering identity in early childhood. *American Journal of Engineering Education, 6*(2), 61–68.

Petroski, H. (2016, September 21). Refractions: Feeling superior? *PRISM*.

Portsmore, M., & Milto, E. (2018). Novel engineering in early elementary classrooms. In L. D. English & T. Moore (Eds.), *Early engineering learning* (pp. 203–224). Singapore: Springer.

Potvin, P., & Hasni, A. (2014). Analysis of the decline in interest towards school science and technology from grades 5 through 11. *Journal of Science Education and Technology, 3*, 784–802. https://doi.org/10.1007/s10956-014-9512-x

Purzer, S., & Douglas, K. (2018). Assessing early engineering thinking and deigns competencies in the classroom. In L. D. English & T. Moore (Eds.), *Early engineering learning* (pp. 113–134). Singapore: Springer.

Salado, A., Chowdhury, A. H., & Norton, A. (2018). Systems thinking and mathematical problem solving. *School Science and Mathematics*, 1–10. https://doi.org/10.1111/ssm.12312

Simon, F., & Broadley, L. (illustrator). (2019). *The goat café*. London: Faber and Faber.

Sinclair, N. (2006). The aesthetic sensibilities of mathematicians. In N. Sinclair, D. Pimm, & W. Higgins (Eds.), *Mathematics and the aesthetic: New approaches to an ancient affinity* (pp. 87–104). New York: Springer.

Song, S., & Agogino, A. M. (2004, September 28–October 2). Insights on designers' sketching activities in new product design teams. In *Proceedings of the DETC'04 ASME 2004 design engineering technical conference and computers and information in engineering conference* (pp. 1–10). Salt Lake City, Utah.

Tank, K. M., Moore, T. J., Dorie, B. L., Gajdzik, E., Sanger, M. T., Rynearson, A. M., Mann, E. F. (2018). Engineering in Early Elementary Classrooms Through the Integration of High-Quality Literature, Design, and STEM+C Content. In: English L., Moore T. (eds) *Early Engineering Learning. Early Mathematics Learning and Development.* Springer, Singapore. https://doi.org/10.1007/978-981-10-8621-2_9

Tank, K. M., Rynearson, A., M., & Moore, T. J. (2018). Examining student and teacher Talk within engineering design in kindergarten. *European Journal of STEM Education, 3*(3), Article 10.

Watkins, K., Spencer, J., & Hammer, D. (2014). Examining young students' problem scoping in engineering design. *Journal of Pre-College Engineering Education Research, 4*(1), 43–53. https://doi.org/10.7771/2157-9288.1082

7

STEAM Through Sensory-Based Action-Reaction Learning

Jan Deans and Susan Wright

Perceptual, Sensory and Multi-modal Thinking: The Underpinnings of All STEAM Disciplines

To understand how Science, Technology, Engineering, Arts and Mathematics (STEAM) curriculum content can be applied within the early childhood setting we must begin with focusing attention on how young children learn. As noted by Eisner (2002), the senses are the first avenues to consciousness. It is through the senses that children experience the qualitative world beginning at birth (and even before birth), with initial learning being reflexive and reactive (Piaget, 1953, 1969). For instance, in the early weeks of life, the infant learns to recognize, discriminate and recall, and is motivated to seek out various forms of stimulation and to explore sensorially in a quest to make meaning of everything that is available in the environment. Hence, it is recognized that young

J. Deans (✉) • S. Wright
The University of Melbourne, Melbourne, VIC, Australia
e-mail: j.deans@unimelb.edu.au; susan.wright@unimelb.edu.au

C. Cohrssen, S. Garvis (eds.), *Embedding STEAM in Early Childhood Education and Care*, https://doi.org/10.1007/978-3-030-65624-9_7

children's learning is based on sensory explorations and relies on active and physical engagement with the world, as is illustrated in the anecdote 'Pink Pig'.

Pink Pig

It is bath time for Alfie. Upon contact with the water, he begins to move his arms rhythmically up and down splashing the water. He focuses intently on the impact that his physical actions are creating; he is watching the droplets rise and fall and the surface patterns created by the moving water. He is exploring through play the properties of water, enjoying its soothing texture, its translucent quality and most importantly his capacity to displace the water with his hands and create an intriguing visual of droplets and patterns.

Alfie notices a rubber pink pig resting on the edge of the bath. He is determined to seize it and reaches out, turning his whole body towards the object, stretching to grasp the toy in both hands. He immediately puts it in his mouth, chews on it, turns it over, and passes it from one hand to the other, squealing with delight as he smashes it through the surface of the water several times. He focuses intently on the object's capacity to displace the water. He repeats the action over and then by chance, he squeezes the pink pig and a stream of water hits him in the face. This is exciting! He registers surprise and for a moment he looks as though he might cry, but he once again thrusts the pig down into the water and draws it upwards squeezing it again, watching the stream of water reappear. Alfie is engaged in sensory-based action-reaction play that is supporting his growing understandings of the properties of water, cause and effect, the impact of bodily force on materials, and his position in space.

Eisner (2002) describes cognition as a generic process of coming to know the world through the senses. This example of Alfie's learning exemplifies how the sensory system stimulates thinking and cultivates imaginative and expressive capabilities. Thus, it is recognized that the learning process for the young child is dependent on a host of perceptual and information-processing activities. These activities stimulate the abstraction of a wide range of sensory-based concepts, which are made available from the environment (e.g. 'warm', 'patterned' water; 'floating/sinking' pig; 'squeezing' to produce an outcome; 'spatially' locating the pig in relation to self; 'bodily' movement to reach the pig; repetition of 'action'). Through sensory-based encounters, concepts are being formed, reformed and then joined with other concepts (Piaget, 1969; Taylor, 1988). For instance, Alfie discovered that the pig was hollow, which allowed water to enter it. Squeezing the pig caused the water to squirt out. Such conceptual learning enables children, even when they are 'preverbal', to make meaning of their environment, predict probability patterns and regulate a wide range of interactions with the world and people in it.

Young children are acutely 'sensorially tuned in' to their immediate environments, absorbing wide-ranging stimuli through the senses of sight, hearing, smell, taste and touch (Hull & Nelson, 2005; Narey, 2009). Children's kinaesthetic responses enable actions to be brought upon objects that they encounter. Hence, learning for young children can be explained in terms of 'sensory-based action-reaction' combined with innate curiosity, which leads to the exploration of what objects look like, what they feel like, how they move and what can be done with them. As with the anecdote of Alfie, you only need to watch the young child at play to recognize how important the senses are when it comes to learning: the child touches an object, registers whether it is hard, soft or malleable, whether it can be grasped, sucked or stretched, whether it can be rolled tipped or pushed, whether it is capable of making sound and whether it is pleasing to the taste buds. Thus, we can appreciate that the child's intuitive understanding of the world is based on actions and reactions to a rich array of environmental stimuli.

Montessori, a renowned early childhood theorist, drew attention to the important link between thinking and physical activity. She noted that intellectual development is inextricably connected with movement and, indeed, is *dependent* on it. Piaget (1969) also identified that perception and motor activity are at the core of young children's thinking, and that young children imitate actions and come to understand that objects are autonomous and independent of self. In the process, their thinking and feeling becomes increasingly systematic and well organized. In addition, children begin to produce new behaviours and novel events which fall within the realm of creative and artistic experience. Symbolic thought emerges as a natural outcome of natural play behaviours all of which rely on multi-modal explorations of a multi-modal world. Indeed, STEAM can be recognized as multi-modal 'linked symbol systems' (e.g. mathematical, scientific, artistic) that support integrated and holistic learning (e.g. visual, spatial, temporal, physical, numerical, written). For young children, such holistic learning comes easily, and children often unite symbol systems through play or playfulness (e.g. singing while painting, dancing to the rhythm of environmental sound or music). Governed by innate natural curiosity and a desire to explore and learn about the world, children demonstrate a significant capacity to live in and through their playing. Play is consequently acknowledged to be a creative act and early childhood scholars recognize that it is through play that much of children's learning is achieved (Fleer, 2017; Grieshaber & McArdle, 2010; Niland, 2009).

Through play, children's spontaneous and improvisational processing of ideas, forms, colours, shapes, materials, sounds and movements can lead to new interpretations or impressions. As such, children have ongoing opportunities to 'self-scaffold' their learning, and to enliven and enlarge their imaginations, as was the case with Alfie in self-initiated play with the pink pig. In the processing of experience, young children find out what they need to "know next through self-initiated problem-solving/finding" (St. John, cited in Connery, John-Steiner, & Marjanović-Shane, 2010, p. 66).

In his writings about children's imagination and creativity, Vygotsky (1978) introduced the notion of the *circular path of imagination*. This process involves the filtering of lived experience through imaginative embodied processing, which combines and recombines elements of the experience, to create artistic products such as an image, music, dance or story.

By way of another example, Alfie has a strong attachment to his cheesecloth swaddling blanket (his 'Shnugie'). When he requires comfort, he seeks it out either nuzzling his face into it or draping it over his entire body. The Shnugie provides emotional security and maintains Alfie's primal-level connection to his mother through association with earlier experiences during the significant attachment period of development. But the Shnugie also offers many opportunities for play, with

Alfie using it for hiding, playing peek-a-boo and for draping over his toys or his mother. He is observed covering his whole body with the material, dragging it slowly off and draping it over his toys that he is pushing around in his trolley. Such play behaviours demonstrate how the circular pathway of imagination is activated through sensory-based action-reaction learning and how this learning provides opportunities for the integration of STEAM disciplines. For instance, Alfie explores the concept of three dimensionality by draping the Shnugie over his whole body; 'becoming' the form of an enclosed shape in space. He also explores concepts such as light and dark, and open and closed. Dramatic intent is also embedded in his play, as he enacts 'surprise' (similar to peek-a-boo), with mother as play partner and audience. This surprise moment provides an extra dimension to his artistic explorations. According to Vygotsky, for the circular path to be completed, intellectual, emotional, physical and social factors need to be involved. As exemplified, Alfie has established a relationship between creativity and multi-modal experiencing, with playful enactment being fundamental to the experience.

Through play, the young child transforms information from the material and social world of lived experience. In this process, the young child learns about everyday concepts through repeated or spontaneous experiences. For instance, through the bathing experience, Alfie explored the material concepts of size, shape, form, enclosure, under, over and surrounding; he also learned social concepts of reciprocal play with his mother: child as initiator/mother as responder.

It is through open-ended play and language exchange such as the play we have described that young children develop and re-conceptualize these everyday concepts into more formal scientific concepts (e.g. through other experiences with water play, children might encounter, for instance, concepts about heavy/light, floating/sinking or still/flowing). Children's transformations from everyday concepts to scientific understandings are strongly embedded in processes of semiotic meaning-making—which is described in greater detail in the next section.

From Everyday to Scientific Concepts: A Focus on the 'A' in STEAM

In general terms, semiotics is the study of signs of all kinds (a sign being something that stands for something else). Hence, the field of semiotics looks at meanings and messages in a range of forms and across multiple contexts. For instance, *acts* and *objects* are signs that produce meaning, which can be interpreted by others.

This notion strongly underpins the Reggio Emilia-inspired concept of children's "hundred languages" (Malaguzzi, 1998). In early childhood education, semiotics is understood to be the capacity of children to use a variety of symbolic languages—drawing, dancing, painting, singing, dramatizing as examples. Children also learn the multiple ways in which each of these languages might be 'spoken' and 'read'. Through these 'languages', children construct and formulate concepts, and consolidate their knowledge and understandings. Rather than thinking of the discrete disciplines of S, T, E, A and M, a semiotic perspective of an integrated STEAM centres on multiple 'languages' and a range of modes of thinking that work together (e.g. language, movement, imagery). This idea parallels Gardner's (1983, 1993) taxonomy of multiple intelligences, which highlights the capacities of individuals to actively construct understandings across multiple intellectual modes including verbal-linguistic, logical mathematical, spatial visual, bodily-kinaesthetic, musical, interpersonal, intrapersonal, naturalistic, existential and pedagogical intelligences. As the reader will see, neither Malaguzzi nor Gardner talks about discrete disciplines, such as science, technology, engineering, arts and mathematics. Rather, they refer to multiple modalities that cross over several, broad domains and are applicable to a range of disciplines.

Hence, a key goal of this chapter is to illustrate children's integrated learning across a range of disciplines (and their various sign-making systems). For instance, science embraces a sign system that centres around big issues such as time, space, energy and flow. Indeed, these constructs are also fundamental to dance (Laban, 1963), music and drama. Likewise, the disciplines of technology, engineering and mathematics all feature concepts of time, space, energy and flow, albeit in different ways. Each discipline has its own distinct symbol system, with its own 'affordances' of how it might be used. In other words, sometimes it's clearer to communicate through numbers than through words, through movement than through sound, or through music than through poetry. Humans, of all ages, learn to select the system that is most suited for the expression of their meaning. Young children, for instance, might at times find it easier to draw a picture about an experience than to give a narrative account of the event—at other times it might be easier to talk rather than illustrate.

Hence, each of the sign systems has unique organizational principles that have elements and conventions that do not have precisely equivalent meanings within other sign systems. For instance, the symbol system of the visual arts centres on principles surrounding the use of colour, shape, texture, composition and the like. There are no direct equivalents within the symbol system of mathematics or language, which use constructs such as numbers, formulae, letters or words. Yet 'equivalency' can function at a metaphoric level, through 'likenesses'. By way of example, in poetry, words function similarly to music, where the rhythm, timing and intonation of the words create type of energy and flow. Likewise, mathematical ideas may be used to shape the design of a pattern, or visual perspective within an artwork—matters that have to do with space.

Young children seem to quite naturally grasp such metaphoric qualities of meaning-making and readily bridge symbol systems to more accurately communicate that which may not easily be communicated through a single mode. Observers of young children will note that children cross modalities and bring their funds of knowledge to new learning experiences. In the example below, Ryle immediately recognizes the affordances of aesthetically stimulating fabric to enact the process of metamorphosis, through costuming, adornment and dramatization.

Ryle has been engaged in a unit of learning about metamorphosis. The content has focused on caterpillars, growth, cocoons, transformation, and butterflies. He chooses to explore the transformation from caterpillar to butterfly through dance. He is attracted to the colour and quality of a vibrant yellow, diaphanous material with the express purpose of enhancing his dance exploration. He gathers the material around his body moving onto the floor, rolling over and over until he is fully encased. As the music begins, he rolls in the opposite direction, gradually releasing himself from his imagined cocoon. He slowly rises, stretching his arms outwards and upwards, focusing his attention on his very personal transformation from a caterpillar to a butterfly and his emergence from the cocoon—states of 'being' that are beyond his human experience.

Ryle later selected a second 'affordance', drawing, to reflect upon and depict his memory of 'being' throughout the three stages of metamorphosis. Drawing helps him to aesthetically and emotionally construct an alternative expression of his dance experience. His drawing consisted of intertwining circular shapes, and he commented, "I made a circle – It's going to fall down and I'm in there and I died. The big round circle was very heavy, and it fell down." Ryle's comment about 'dying' reflects his understanding of the transformative state of moving from a caterpillar to a butterfly; it also reflects his metaphoric, emotional state of 'being' while shrouded in the fabric and then while emerging and flying in a free condition.

Let us elaborate a bit on what is involved when children 'cross over' various modes. Often this is a multi-modal experience, where one mode mediates another—together, the modes enrich and inform each other as the child engages in processes of meaning-making. Suhor (1984) called this 'transmediation'. Transmediation is central to the sensory-based action-reaction principle we described earlier. In Ryle's case, he transmediated from one concrete reality to an imaginative reality. He drew upon his scientific knowledge of cocoons, his awareness of change occurring within the cocoon, and his imagery of himself as 'being' with wings when he emerges. He transmediates from science concepts into the artistic, imaginative sphere, through dramatic action. Transmediation is a fundamental, perhaps innate, process in young children's thinking and experiencing. When children transmediate, they seamlessly configure and reconfigure their ideas, thoughts and feelings across modes (Kress, 2003). The argument presented in this chapter is that at the *core* of such transmediation is the role of the body and the senses as the child explores, and embodies, a range of concepts. In this example, Ryle is consolidating his learning about metamorphosis through dance and dramatic enactment. He is exploring body shape in space, in time, with energy and flow: in particular, three dimensionality, rising and falling, opening and closing, high and low. These concepts are metaphorically applicable to other disciplines such as music, drama, maths, science and engineering.

The following case study illustrates the process of transmediation through dance-play and drawing-telling. It highlights the link between children's sensory-motor cognition, play and creative imagination and how children transmediate across sign systems as they make meaning. The case highlights the co-existence of scientific (i.e. formal or conceptual) and everyday (i.e. experiential) concepts as children learn. Using teacher-developed "learning stories" (Carr, 2001), photographic and corresponding drawing-tellings (Wright, 2007a, 2007b), the case study demonstrates how children learn STEAM concepts through open-ended playful experiences, multi-modal inquiry and language exchange, and then re-conceptualize these into more formal conceptual understandings.

Within this case study, three modes of children's representation (Bruner, 1966) are featured, namely:

- enactive representation (action-based, in this case, dance),
- iconic representation (image-based, in this case, drawing), and
- symbolic representation (language-based, in this case, narrative telling).

This triad of learning modalities is showcased through an in-depth description of one learning exemplar, to illustrate how multi-modality and transmediation is applicable to a range of STEAM-oriented learning experiences.

A Case Study

The setting was an inner-city university research and demonstration pre-school. The participants were 20 four-year-old children and their teacher who engaged in one-hour weekly sessions in a carpeted multi-purpose open space within the preschool. The ambience was one of 'overall softness', inviting multi-sensory participation. The class structure included a welcome which focused on children articulating their interests and ideas, a physical warm-up, practice of selected dance skills, free dance improvisation, relaxation and drawing-telling.

The topic of butterflies had developed organically out of the children's interest in flying insects and bugs. The teacher facilitated a discussion with the whole group about the life cycle of the butterfly. Visual images of a variety of species of butterflies had been shown and embroidered pieces of lace in the shape of the butterfly had been dispersed amongst the children to enable them to have the hands-on experience of feeling the texture of the lace and to get physically and emotionally in touch with the qualities of lightness, the delicacy of the butterfly wings and butterfly movements (see Fig. 7.1).

Fig. 7.1 Visual and tactile stimuli used to support children's focused participation and creative thinking

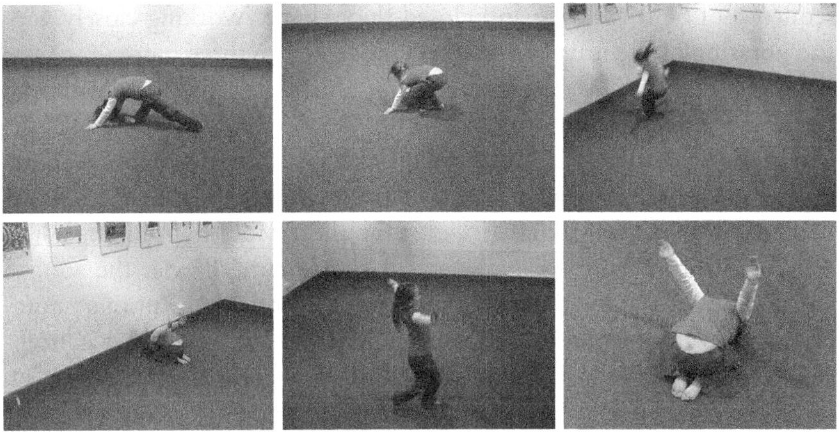

Fig. 7.2 The up and down again butterfly. (Child: Amelee, free dance solo observed by the audience of peers and the teacher, Duration: 58 seconds, Music: Bach Partition Violincelle)

The dance objective was to explore light lifting and lowering of arms, moving through the space lightly, exploring meandering pathways and landing lightly in different spaces in the room. At the time of the introductory discussion, Amelee was recorded as saying: *"I saw hundreds of butterflies at the zoo. The butterflies' wings are paused and open. That's what real butterflies do"*. Figure 7.2 presents still frames of the sequence of Amelee's dance and the teacher's anecdotal recording of Amelee's dance event.

Amelee requests the recorded music used earlier in the class to accompany explorations of light arm gestures. She locates herself in the centre of the room and creates a closed body shape on the floor. She rests on her knees and lower legs, bending her upper body over her thighs. Her head is placed on the floor and her arms are stretched out. As the music begins, she rises and runs quickly and lightly to the back wall facing away from the audience. Working with the rhythmic structure of the music she moves into the floor, closes her torso over her knees and stretches her arms out behind her in a 'wing-like' shape. Amelee pauses in this position and then on cue with the phrasing of the music she rises and returns to a light run, moving centrally forwards in the space towards the audience. She comes to stillness and returns to the body shape on the floor previously adopted. She waits in stillness before rising again. She repeats the sequence of light running and closed floor shape. Working closely with the phrasing of the music she repeats the sequence three more times and with each repeat she is observed smiling. Amelee concludes the dance on the floor returning to the same shape she began with, holding it until the end of the phrasing of the music. She rises and faces the audience smiling.

This record of Amelee's individual dance provides insight into how she drew on her "funds of knowledge" (González, Moll, & Amanti, 2005) obtained from personal experiences of seeing butterflies at the zoo, to engage in a multi-modal exploration using her body as the primary communication tool. Amelee's dance supported her to connect her conceptual understandings with her felt and lived experience. In creating the 'up and down again butterfly dance', she communicated her scientific understanding of flight, together with her mathematical understandings of time and space, as she coordinated her body movement with the rhythms and phrasing within a musical accompaniment. Her thoughtfully selected body activities and gestures, which included a repeated on-the-floor body shape, controlled rising and falling, purposeful use of space and levels, a circular floor pathway, repeated opening and closing arms and light running, enabled her to effectively express a dramatic and aesthetically imbued, free flowing enactment of the butterfly. The centrality of aesthetics was evident in her emotional, cognitive and social learning (Vygotsky, 1978). In this solo performance, Amelee's aesthetic selection of movements coalesced, where her 'internal coherence' became apparent, as properties and parts of her dance became 'a whole'.

The dance also demonstrates Amelee's capacity for symbolization through choreographic form, which is significantly reliant on the processing of mathematical concepts and understandings (Pugh McCutchen, 2006). Through the purposeful selection of movements, spatial structures and repetition, she demonstrates her ability to think kinaesthetically, musically, scientifically and mathematically, integrating the various symbol systems to communicate the particular "dynamic line" (Sheets-Johnstone, 1979) of her artistic thinking processing. Although the dance took place within the genre of improvisation, where the rules vary according to the individual's inclinations, this dance was nevertheless imbued with a considered sense of form and design. Form was exemplified through Amelee's attention to her lifting and lowering arm gestures, and their light quality, which were selected and enacted to communicate her thoughts and feelings, multi-modally.

Amelee commented to the teacher that "*I've put a lot of thought into my dance*", a statement which affirms the role of meta-cognitive processes in her learning. Clearly, she was indicating that she had created a mental model of how the dance would unfold—its movement content, its spatial configurations and its dynamic quality. It might have been that Amelee had devised a choreographic plan for her dance, tried it out, evaluated it, made additions or deletions to it. Her conscious realization of artistic judgments and decision-making were given final form in the dance, where 'thought in action' supported her creative processing.

Amelee crossed dance, graphic and verbal modalities by re-representing, through drawing-telling, her lived experience of observing the butterflies in the zoo and dancing a butterfly in class (Deans & Wright, 2018). Her drawing-telling (see Fig. 7.3) provides insight into her transmediation across symbol systems using visual-spatial imagery and language.

Wright (2007b) explains that "images can be thought of as stopped action frames or visual representations of actions" (p. 2). For instance, the image in Fig. 7.3 presents two stopped action frames, of two moments in time: (a) Amelee located centrally on the paper 'being the butterfly' with arms outstretched; (b) herself again, as a smaller figure to represent the 'down again' movement in contrast with the large movement. As shown

Fig. 7.3 That's me being a butterfly and I went up again and down

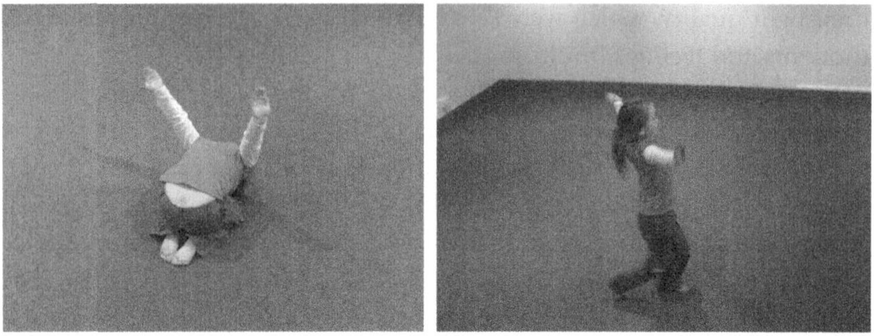

Fig. 7.4 Resting and flying butterfly wings

in Fig. 7.4, the resting and 'flying' image in contrast with the folded floor shape were repeated movement motifs in her dance. Her narrative, "*That's me being a butterfly and I went up again and down*", is intimately linked to the bodily action of her dance.

Wright (2007a, 2007b) noted that, for the young child, moving, drawing and talking are mutually inclusive transformative processes that support the construction of knowledge through *thinking in action*. Brooks (2005) also notes the relationship between drawing and visual thought, proposing that drawing, when "used as a medium of exchange, can form a dynamic function that allows an elaboration of an initial idea and the definition of a concept as well as assisting with building empirical connections between concepts and systems" (p. 5). The inclusion of drawing-telling as a repeated activity at the end of each dance class provided the children with the opportunity to unite memory, imagination and observation. Drawing-telling enabled collaborative discussions between individual children and the teacher, along with the one-on-one process of recording the three key and integrated components of descriptive narrative, expressive vocalisms and gestures (Wright, 2010).

In addition to dancing and drawing-telling, Amelee demonstrated a cross-modal connection between movement and music as complementary semiotic tools for making meaning. The musical recording selected by Amelee was one to which she had danced in previous classes. Amelee embodied the music, coordinating the beginnings and endings of musical phrases with her improvised movement sequences of 'up and down'. In so doing, she was transmediating across the symbolic domains of dance and music, while simultaneously drawing upon inter-related aspects with the disciplines of science (space, force, time, flow), mathematics (above-below, large-small, near-far). She demonstrated her understandings across multiple modes (physical, aural, visual) and self-scaffolded, seamlessly interchanging a number of semiotic tools for the express purpose of artistic meaning-making.

These are indications of Amelee's capacity for abstract thinking, evidenced through her manipulation of structural-mathematical dance elements, including spatial forms, levels, variety, contrast, harmony, repetition, phrasing and sequencing—all of which were intentionally selected to give the dance its communicative form. Amelee was not attempting to 'express herself' per se, but rather, she was using her scientific and mathematical knowledge to communicate/express her personalized feelings about the delicate *quality* of the butterfly and the *lightness* of its movements. Her dance was an outcome of imaginative and affective consciousness (Wright & Deans, 2020a) that was given form through the selection and manipulation of the discipline knowledge of dance, music, mathematics and science.

Amelee was clearly captivated by the movement and dynamic qualities inherent in the butterflies that she had observed at the zoo; an image that was further stimulated by her holding and moving the delicate lace butterfly that her teacher had presented to the group as a sensorial stimulus (see Fig. 7.1). These images triggered her desire to engage in a creative process, which enabled her to integrate a number of symbolic forms in a purposeful manner, dramatically enacting a multi-modal response. The 'up and down again butterfly' solo tells the observer a lot about four-year-old Amelee's knowledge and capacity for symbolic, expressive communication. Amelee represented what she had experienced in nature, which required her to recall visual information and to use her imagination and body to metaphorically recreate the qualities of 'butterflyness'. Through the enactive mode of dance, and the iconic/symbolic mode of drawing-telling, Amelee symbolized her felt understanding, by *assembling* the following:

- perceptions of sensations from the environment,
- imagined ideas, thoughts, images and emotions, and
- embodied time, space, force and flow (Wright & Deans, 2020b).

The resultant artistic outcome is technically sound and filled with high levels of emotional content that is sensitively communicated through the integration of body, mind and spirit—a state that is identified as *perezhivanie*, or intensely emotional lived experience (Vygotsky, 1978). At no time throughout the dance did she lose her aesthetic sensibility for the re-enactment of the butterfly; it seemed she felt internally the intensity and imagery which was supporting her self-initiated ZPD, leading her to challenge herself to perform at a level that seemed like she was a "head taller" (Vygotsky, 1978)—in other words, instantly, more 'developed'.

In summary, this analysis and synthesis of Amelee's 'up and down again butterfly' solo provides insight into how dance and drawing-telling stimulates STEAM thinking and learning, through a particular emphasis on perception, imagination and embodiment, rather than a narrow focus on specific disciplinary areas. For Amelee, the coordinated dance to music (and ensuing drawing-telling) moved her understandings beyond the everyday and predictable, into a space where aesthetic reasoning and emotional connectedness resulted, and where the communication of novel forms of body thinking and expressing were enacted and acknowledged. The open-ended dance-play improvisation and experimentation provided her with a natural way of learning about the world and about how to depict this world via movement and via pen-on-paper. It also offered her a dynamic and kinaesthetically oriented opportunity to access her imagination, to make concrete the familiar and to combine known understandings with new ways of experiencing. In short, to see knowledge in a new light.

Such processes are akin to those used in creative engagement within the fields of mathematics and science. Indeed, as illustrated through Amelee's exemplar, dance, drawing, mathematics and science all involve processes of critical analysis and reflection, re-creation and re-formation. As with all forms of higher order thinking, we transform our everyday concepts (e.g. butterfly) with our thinking, feeling and aesthetic understandings and scientific concepts (e.g. metamorphosis). When this happens, we often experience what Vygotsky (1978) terms catharsis, or in other words, emotional knowledge of an embodied image or concept. As was the case with Amelee, such encounters bring into focus a range of learning dispositions such as motivation, self-control, self-determination, persistence, meta-cognition, imagination and creative innovation. In turn, these dispositions stimulate higher order thinking, often blending body, mind and spirit.

Conclusion

This chapter has discussed young children's learning from the perspective of sensory-based action-reaction meaning-making, which is multi-modal and reliant on the social construction of knowledge. STEAM is presented as an integrated curriculum form that relies on transmediation across modes to enable flexible and holistic learning. The exemplars of Alfie, Ryle and Amelee highlight how artistic, expressive explorations play a central role in supporting generative thinking and learning, through multiple sign systems, each of which has unique organizational principles. When these sign systems work together, they afford a deep level of engagement in thinking, feeling and learning processes that transmediate across multiple modalities. The case studies also exemplify how young children engage in problem solving, decision-making and creative thinking, as established connections are reinforced, and new understandings developed across various disciplines.

Takeaway Message

Teachers and parents have an opportunity to consider new possibilities for teaching and learning STEAM by providing opportunities for children to make meaning across multiple modes, rather than considering the discipline areas singly. As Kress (2003) reminds us, young children learn about their world through multiple sign systems long before they are exposed to separate subject areas. This chapter stimulates readers to think about sensory-based action-reaction thinking and experiencing and the important role that adults play in helping young children to fully explore STEAM concepts and ideas through multi-modal meaning-making.

References

Brooks, M. (2005). Drawing as a unique mental development tool for young children: Interpersonal and intrapersonal dialogues. *Contemporary Issues in Early Childhood, 6*(1), 80–91.

Bruner, J. S. (1966). *Toward a theory of instruction.* Cambridge, MA: Harvard University Press.

Carr, M. (2001). *Assessment in early childhood settings: Learning stories.* London: Sage Publications.

Connery, C., John-Steiner, V. P., & Marjanović-Shane, A. (2010). *Vygotsky and creativity. A cultural-historical approach to play, meaning-making, and the arts.* New York: Peter Lang Publishing.

Deans, J., & Wright, S. (2018). *Dance-play and drawing telling as semiotic tools for young children's learning.* London: Routledge.

Eisner, E. (2002). *The arts and the creation of the mind.* New Haven: Yale University Press.

Fleer, M. (2017). *Play in the early years* (2nd ed.). Port Melbourne, VIC: Cambridge University Press.

Gardner, H. (1983). *Frames of mind: The theory of multiple intelligences.* New York: Basic Books.

Gardner, H. (1993). *Multiple intelligences: The theory in practice.* New York: Basic Books.

González, L., Moll, C., & Amanti, C. (2005). *Funds of knowledge: Theorizing practices in households, communities and classrooms.* Mahwah, NJ: Lawrence Erlbaum.

Grieshaber, S., & McArdle, F. (2010). *The trouble with play.* Berkshire UK: Open University Press.

Hull, G., & Nelson, M. (2005). Locating the semiotic power of multimodality. *Written Communication, 22*(2), 224–261.

Kress, G. (2003). *Literacy in the new media age.* London: Routledge, Taylor Francis Group.

Laban, R. (1963). *Modern educational dance* (Rev. 2nd ed.). L. Ullmann (Ed.). London: MacDonald and Evans.

Malaguzzi, L. (1998). History, ideas and basic philosophy: An interview with Lella Gandini by Loris Malaguzzi. In C. Edwards, L. Gandini, & G. Forman (Eds.), *The hundred languages of children: Advanced reflections.* London: Ablex Publishers.

Narey, N. (Ed.). (2009). *Making meaning. Constructing multimodal perspectives of language, literacy, and learning through arts-based early childhood education*. Pittsburgh, USA: Springer.

Niland, A. (2009). The power of musical play: The value of play-based, child-centered curriculum in early childhood music education. *General Music Today, 23*(1), 17–21.

Piaget, J. (1953). *The origin of intelligence in children*. New York: Int. University Press.

Piaget, J. (1969). *The mechanisms of perception*. London: Routledge & Kegan Paul.

Pugh McCutchen, B. (2006). *Teaching dance as art in education*. Chelsea, MI: Sheridan Books.

Sheets-Johnstone, M. (1979). *The phenomenology of dance*. London: Dance Books Ltd.

Suhor, C. (1984). Towards a semiotic-based curriculum. *Journal of Curriculum Studies, 16*(3), 247–257.

Taylor, M. (1988). The development of children's understanding of the seeing-knowing distinction. In J. W. Astington, P. L. Harris, & D. R. Olson (Eds.), *Developing theories of mind* (pp. 207–225). New York: Cambridge University Press.

Vygotsky, L. (1978). *Mind in society. The development of higher psychological processes*. Cambridge, MA: Harvard University Press.

Wright, S. (2007a). Graphic-narrative play: Young children's authoring through drawing and telling. *International Journal of Education and the Arts, 8*, 1–28.

Wright, S. (2007b). Young children's meaning-making through drawing and 'telling': Analogies to filmic textual features. *Australian Journal of Early Childhood, 32*(4), 37–48.

Wright, S. (2010). *Understanding creativity in early childhood: Meaning-making and Children's drawing*. Washington DC: Sage.

Wright, S., & Deans, J. (2020a). Learning through the Arts. In A. Kilderry & B. Raban (Eds.), *Strong foundations. Evidence-informed Practice for early childhood*. Camberwell, VIC: ACER Press.

Wright, S., & Deans, J. (2020b). The role of movement, dynamics and expression in children's drawings of dancing. In C. S. Nielsen & S. Burridge (Eds.), *Dancing across borders: Perspectives on dance, young people and change* (pp. 157–169).

8

To STEAM or Not to STEAM: Investigating Arts Immersion to Support Children's Learning

Susan Narelle Chapman, Georgina Barton, and Susanne Garvis

Introduction

Over recent years the concepts of interdisciplinary learning involving Science, Technology, Engineering and Mathematics (STEM) and additionally Science, Technology, Engineering, Arts and Mathematics (STEAM) education have become popular approaches in a range of educational contexts, especially within the early years. Despite this

S. N. Chapman (✉)
Queensland University of Technology, Brisbane, QLD, Australia
e-mail: s23.chapman@qut.edu.au

G. Barton
University of Southern Queensland, Darling Heights, QLD, Australia
e-mail: Georgina.Barton@usq.edu.au

S. Garvis
Swinburne University of Technology, Hawthorn, VIC, Australia
e-mail: sgarvis@swin.edu.au

C. Cohrssen, S. Garvis (eds.), *Embedding STEAM in Early Childhood Education and Care*, https://doi.org/10.1007/978-3-030-65624-9_8

popularity, there is limited evidence in the scholarly literature highlighting the benefits and importance of STEAM for children's learning outcomes and experiences. Further, some reports outline how the arts (dance, drama, media arts, music and visual arts) are often superficially tagged onto a STEM approach rather than embedding the arts authentically through learning and teaching.

Advocates for a STEAM approach suggest scientific progress will be enhanced by the type of creative thinking that is embedded in the arts (Daugherty, 2013; Taylor, 2016). The concepts which underpin STEAM have been traced back to the ideas of John Dewey, who recommended an integrated curriculum with real-world applications to improve students' practical understanding of the world (Watson & Watson, 2013). Supporters of interdisciplinary learning hold the view that this holistic approach to the curriculum can connect interdisciplinary knowledge and provide students with a new perspective on their "home discipline" (Guyotte, Sochacka, Costantino, Kellam, & Walther, 2015). There is a general acceptance in most of the STEAM literature that interdisciplinarity is a way forward, though how that could happen is often not clear. In this chapter we demonstrate how this could happen.

Enacting a STEAM (rather than STEM) approach allows children to become more active learners and provides natural opportunities for exploration. The arts provide children with other ways of knowing and understanding (Eisner, 2003, 2005), making meaning by using their bodies and their minds together so that thought, emotion and action are blended (Wright, 2012). Young children naturally tend to connect the mind and the body as equal partners in learning (Schiller & Meiners, 2012). By learning in, with and through the arts, young children can use uniquely expressive and symbolic ways to communicate what they know (Wright, 2012). These ways of learning differ from scientific ways of learning. While there is much talk calling for STEM learning areas to embrace creative learning strategies (Taylor, 2016), Arts-based approaches to teaching and learning provide workable opportunities for this to happen (Daugherty, 2013). However, arts-based approaches require teachers and families to have adequate knowledge and skills in these areas.

Critics of integrated learning approaches—like STEAM—have been concerned that this approach will lower teaching and learning standards by omitting the development of knowledge and skills that are associated with specific subjects (Humes, 2013). There are reports that the subjects that are included in the STEM group have often been taught in traditional ways that concentrate on students repeating what they are told and sitting at their desks to learn by listening and writing (Taylor, 2016). Concerns have been expressed that these approaches to teaching older students have been reflected in the increased emphasis on formal schooling for children as young as three years of age (Ang, 2014). This view has been supported by the 'back to basics' motto which tends to favour teaching to a formula and rote learning. By contrast, the creative and expressive capacity of the arts can help children to view science as a creative human endeavour (ACARA, 2015). It does not follow that integrated learning negates the need for subject-specific knowledge, or that links to subject-specific curricula are ignored. High-quality integrated learning embraces both the intrinsic value of each subject *and* the bigger picture thinking that can connect different ways of thinking.

With the influence of high-stakes testing in schools, teachers may feel like "the meat in the sandwich" (Klenowski, 2011; Thompson, 2013). The benefits of wide-ranging creative teaching strategies may not be included in their pre-service training and cannot be guaranteed in their practicum learning placement. Faced with the demands of ensuring children achieve high test scores, teachers may either be unaware of creative teaching strategies or feel pressured to ignore creative approaches and teach to the test (Garvis, 2012; Garvis & Lemon, 2013). Some children do not prosper in such learning environments, but children are able to thrive when educators acknowledge that learners are not all the same and one size does not fit all. Teachers and families can play an important role together in helping all children to learn in the most effective ways possible. The opportunity to learn well needs to be offered to all children, and not delivered in a way that favours some children over others. Research suggests that when teachers use creative arts-based approaches to learning, rather than organizing their teaching around practising for high-stakes tests, students are more engaged in learning, retain their learning longer and apply knowledge and skills to other contexts independently of teacher instruction (Chapman, 2018). The arts thus play an important role in unlocking learning for children.

In this chapter we describe an Arts Immersion approach in classrooms to support improved learning outcomes (Chapman, 2018) and to optimize the use of children's first languages: play and the arts (Wright, 2012). This approach could be adopted to support learning across all ages and year levels. Arts Immersion therefore pushes back against the dominance of word-based text in the classroom, which may privilege some students and disengage others (Chapman, 2020) and advocates for "the use of any form of representation in which meaning is conveyed or construed" (Eisner, 2003, p. 342). As the arts represent a unique set of languages, they are used in the case studies described below as a domain of learning (learning *about* the arts) and as a vehicle to access other learning (learning *with* and *through* the arts). Arts Immersion acknowledges children's diversity and offers more inclusive strategies for learning and teaching. We suggest this approach is highly beneficial for teachers to support the learning of all children—from those who benefit from additional support to those who have mastered core ideas and are eager to learn more. The interdisciplinary use of these arts languages in educational contexts creates opportunities for working collaboratively with STEM subjects and reshaping children's experiences of learning. Early years curricula have an advantage in this regard since these documents describe outcomes which are not subject-based and provide opportunities for the seamless involvement of arts experiences for children. The curriculum focus is on play-based learning and communication as well as social and emotional development. These align effectively with arts-based methods and enable learning and teaching in early childhood settings to be more effectively integrated across the curriculum, rather than siloed into separate subject areas.

Arts Immersion Approach

Arts Immersion refers to the purposeful use of the arts to support learning across different disciplines (Chapman, 2015). STEAM, which integrates the arts and STEM learning areas, sits within an Arts Immersion approach, and involves teaching with and through the arts, *across* the curriculum. Early years curricula align well with an Arts Immersion approach and integrated STEAM learning. For example, in the Early Years Learning Framework for Australia (Department of Education Employment and Workplace Relations, 2009), children are explicitly encouraged to develop learning dispositions which include curiosity, creativity and imagination, and to express ideas and make meaning using a range of media. In some cases, due to limiting interpretations of national curricula, a hierarchy of teaching strategies may emerge where creative approaches are regarded as less worthy. The case study examples in this chapter explore how a broader arts-based pedagogy can benefit learning and teaching for young children. The findings from the case studies also suggest that the enactment of STEAM education, through learning and teaching approaches common to early years contexts, may hold valuable lessons for teachers across all year levels.

An Arts Immersion approach to learning and teaching aligns with embodied learning in which the body and the mind work together closely to engage the students' senses and intellect (Wright, 2012). As the intent of teaching is to develop learning, embodied pedagogy concerns the relationship between teachers and their students, and between their teaching and learning practices. From this viewpoint, our minds and bodies unite in helping us to interpret our experiences and perceptions of the world (Stolz, 2015). Interdisciplinary learning with and through the arts lays the foundation for broader learning choices and a greater variety of teaching strategies that encourage active learning. Knowledge is constructed through embodied learning experiences rather than being acquired only through instruction.

However, arts-based pedagogies may be less popular in primary schools than in early childhood education settings. For example, the EYLF (DEEWR, 2009) supports a natural interdisciplinary approach (Goodfellow, 2009). The curriculum for older students, however, is arranged in distinctly separate learning areas (Daugherty, 2013; Ewing, Miller, & Locke, 2014). For example, the EYLF proposes that creative and expressive arts are ways for children to express ideas and make meaning, leading to their development as effective communicators. Furthermore, early childhood teachers are encouraged to use a wide range of teaching strategies to accommodate different ways of learning. However, this approach does not appear to be emphasized in the interpretation of curricula for older students, and concerns have been expressed regarding the wisdom of this approach when children start formal schooling (Humes, 2013). As authors, we suggest that interdisciplinary approaches should be continued into formal schooling to support the learning of all children.

Enacting an Arts Immersion Approach to STEAM

Two case studies are presented to demonstrate the benefits of authentic STEAM approaches to learning and teaching. It is hoped they will inspire our readers to enhance their own interdisciplinary pedagogy across all levels of education in implementing STEAM approaches. Ways to extend STEAM education into the home environment by engaging families in their children's learning are also suggested.

Case Study 1: Bird Feeders

Infants and toddlers instinctively explore the world through their senses, making meaning through experiences that are first experienced in their bodies before they are processed in their minds. Infants and toddlers express their knowing by enacting what they understand. As such, the arts function as natural interdisciplinary languages that transcend disciplinary, cultural and developmental boundaries.

In this case study, Arts Immersion was chosen by the teacher in a toddlers' room to explore a design approach based on the children's interest in birds. The teacher had extensive experience with Arts Immersion approaches and encouraged children to engage in sensory exploration. The researcher documented the teacher's daily interactions using written observations and photographs. In addition, the teacher regularly discussed her approach—and the rationale for her approach—with the researcher to explain specific planned learning experiences.

The teacher drew on developmental pedagogy as her learning framework (Pramling Samuelsson & Asplund Carlsson, 2008) Developmental pedagogy includes specific features of dialogue (e.g. the use of questioning with children), a focus on learning acts as opposed to learning objects (e.g. what do I want the children to do today, rather than what do I want the children to develop in terms of arts ability), and discernment/variation to enhance children's learning (e.g. creating variation in an activity such as different tempos in music or different shades of the same colour). According to Pramling Samuelsson, Asplund Carlsson, Olsson, Pramling, and Wallerstedt, "to learn means to change from one way of experiencing something to another way of experiencing the same thing" (2009, p. 124). Accordingly, the role of the early years teacher in this approach is to create opportunities, situations and activities that challenge children's ways of experiencing or making sense of something (Pramling Samuelsson & Asplund Carlsson, 2008).

The point of departure for the teacher in this case study was the children's interest in the birds that had recently returned to the trees in the playground. The children were excited to greet the birds each day when they went outside. They wanted to know more about how birds lived and what they ate. Initially, the teacher asked the children to draw and paint the birds they had seen outside. Next, the teacher taught the children songs about birds and filled the room with various books about birds and images of birds. The goal was to show the children how birds lived and ate. Videos about birds were also incorporated into circle time to prompt conversations. During one circle time, the children began to talk about building a bird feeder they had seen in a video. As children's verbal language skills were emerging, the teacher encouraged the children to express their ideas in drawing and artwork. This enabled the teacher to develop an understanding of the children's perspectives and to understand the children's sense-making (Doverborg & Pramling Samuelsson, 2000). The teacher asked open-ended questions to encourage the children to share their ideas and express their understanding. Further, the children were also invited to ask questions. In this way, children's knowledge and interest guided the learning process.

The children began to design their bird feeders on paper, before working with recyclable material to build the design. The children engaged in various drawings and paintings, experimenting with different visual arts materials in their design. During this process, the teacher spoke about different shapes and incorporated various objects into art activities. The teacher had two different size recyclable material objects for the children to explore with a toy bird and the children consequently decided that milk containers were a good size for small birds after experimenting to see which objects the toy bird would fit into.

The children had learnt from watching a video that some birds are attracted to bright colours. Consequently, the children decided to paint the bird feeders in bright colours so the birds could see them. Their teacher continued to ask questions to encourage children's learning about the colour, shape, texture and design of the bird feeders as well as about painting techniques. Conversations were organic and responsive to each child over time, guided by children's individual understandings about the bird feeder.

When the bird feeders were finished, the children put them in the garden. At the end of the week, the children took their bird feeders home. Parents had been informed about the STEAM learning journey; they were encouraged to use the bird feeder at home and to continue conversations about birds. Parents were also asked to encourage their children to keep a record of the birds (in paintings and drawings) and over the following week, to share this with the teacher when the children came to preschool. With the teacher's support, the toddlers kept a tally of the number and colour of the birds they had seen at home using the children's drawings of birds (in the correct colour) as tally marks. These were added to the tally chart and thus constituted data that could be used to answer simple questions, such as 'Were there more brown birds, or black birds?' In this way, continuity of learning across settings was supported, families were included in the early learning centre's curriculum development and meaningful conversations between teacher, families and children were supported.

Case Study 2: Electrons

An Arts Immersion approach is also effective in primary school. In this case study, we demonstrate how it was enacted with school students who had highly diverse backgrounds in the context of a research project. A teacher who was interested in using arts-based pedagogy agreed to take part in a project that focused on science—specifically, learning about electrons. Both the school principal and the participating teacher had extensive experience in early childhood education. However, the participating teacher was new to primary school teaching.

Critical Participatory Action Research (CPAR) was employed to explore how using the arts, as a core approach (*Arts Immersion*), might influence learning and teaching with primary school students. CPAR involves investigating practices, such as a teaching practice, with the intention of bringing about improved outcomes for the participants (Kemmis, McTaggart, & Nixon, 2014). In addition, the theoretical lens of practice architectures (Kemmis et al., 2014) was used to frame researcher and teacher observations and reflections as 'sayings' (thinking space), 'doings' (physical space) and 'relatings' (social space). These theoretical lenses were applied to understand the ways in which students communicated their understanding, used their learning space and resources and related to each other and to the teacher and researcher.

During the data collection period, the researcher worked alongside the teacher. First, the teacher provided her lesson plans to the researcher. The researcher developed arts experiences that supported learning and assessment and were aligned with the lessons. In a second step, the researcher and the teacher refined the arts experiences before the teacher used them in class. As often happens, timetable changes required flexibility. A Year 3 teacher at the same school was interested in this process. She volunteered to observe some of the classes and to provide feedback—both informally and through interviews.

A preliminary trial gave the researcher an opportunity to build a collegial relationship with the teacher and to get to know the students. Thereafter, the process of lesson planning, Arts Immersion activity planning, refining and classroom delivery was repeated four times. We called this phase, the 'Reconnaissance' stage. The purpose of the Reconnaissance stage was to gather preliminary information and to introduce Arts Immersion as a new approach since the arts were usually taught as a separate subject.

The activity involved a collaboration between drama and science in which students enacted the flow of electrons moving in a negative-to-positive direction, from a battery (the power source) to a light bulb. The battery and light bulbs were represented by green buckets with picture labels. The 'battery' bucket was filled with pegs which symbolized small quantities of energy. Embodying and communicating learning, the students picked up the pegs from the 'battery' bucket and carried them in the appropriate direction to the 'light bulb' bucket where they deposited the pegs, indicating that the electricity had been transferred from the battery to the light bulb. The reconfiguring of furniture in the classroom provided a large space that was suitable for accommodating movement and not defined by desks. The teacher and researcher observed how the students used their learning space and resources, and how students related to each other and to the teacher and researcher.

Students engaged in coordinated movement, were energized, and enjoyed working with their peers in the reconfigured space to represent an electrical circuit. This simple collaborative activity was focused on direction of movement rather than expressive quality, although students chose to modify this movement to suggest a 'robotic' and continuous flow of energy without gaps or interruptions. Students were able to transfer their bodily kinaesthetic understanding to drawn diagrams constructed immediately straight after this drama-based activity (see Fig. 8.1).

By first enacting the movement of electrons in an electrical circuit and embodying this knowledge, students were able to retain their learning and express it in other forms.

> *The drama that was used to illustrate the movement of electrons made a huge impact as reflected in student assessment when they drew the drama.* (Researcher Journal) (Chapman, 2018)

Figure 8.2 presents an example of formative assessment gathered later in the week. The use of drama to enact the movement of electrons had reinforced Heidi's learning.

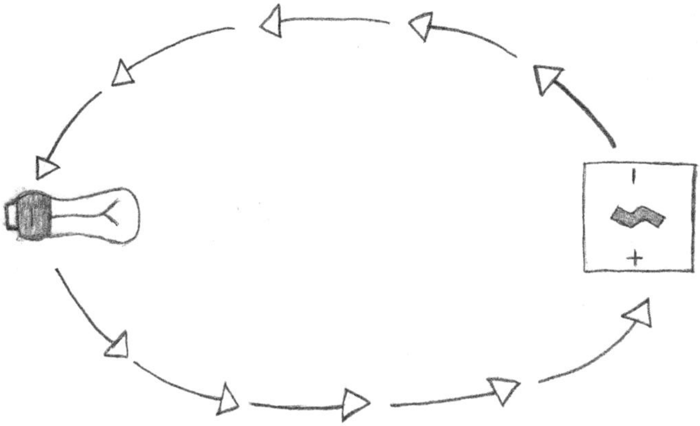

Fig. 8.1 Electrical circuit

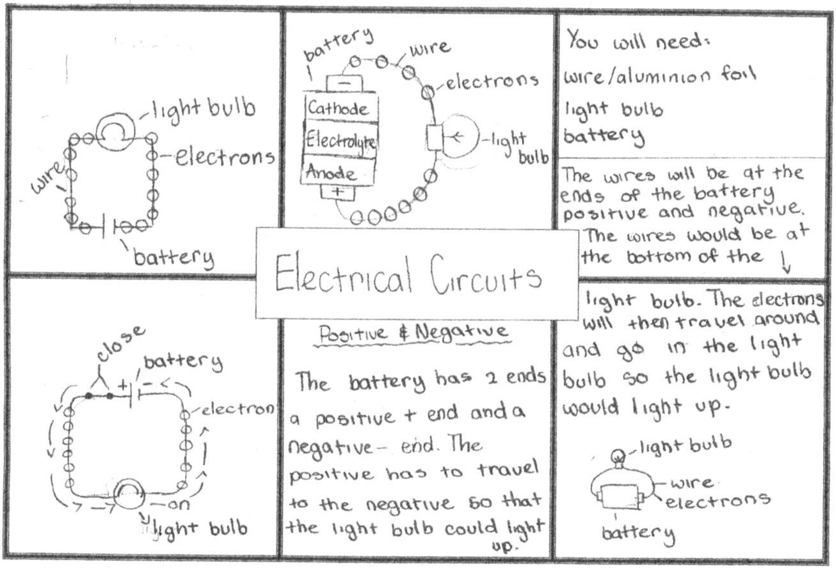

Fig. 8.2 Heidi's formative assessment response. (Chapman, 2018, p. 85)

Students were welcome to provide their written, formative assessment responses using word- and picture-based representations of learning. This was in marked contrast with established practice at the school. Pictorial representations can also act as a stimulus for discussion equipping students to refer to the pictures as a starting point for explaining aloud.

This simple activity fostered the connection between mind and body. It included all students, regardless of developmental level, disability, culture, gender or language background. Language learning in the primary years may often be regarded as synonymous with word-based text. However, in this STEAM approach, the arts language of drama became the language of learning about the flow of electrons in an electrical circuit. Rather than limiting learning and lowering achievement standards, this STEAM activity promoted clearer understanding and helped students to remember what they had learned.

In the course of this research project, students became 'arts ready': their familiarity with the use of arts languages across the curriculum was established. Over time, Arts Immersion activities, including STEAM strategies for learning, grew in complexity. Students who did not usually engage with STEM subjects or attain required achievement levels were supported in reaching improved outcomes. By working with a researcher who was an arts specialist, professional learning occurred in the classroom. The opportunity for the teacher to learn 'on-the-job' provided an effective strategy for building teacher capacity in using an integrated STEAM pedagogy, as part of a wider Arts Immersion approach to learning and teaching.

Moving Forward

An arts-based approach to teaching creates opportunities for children to share their knowledge and ideas in creative ways. STEAM, as one aspect of an arts-based approach, draws on the arts to enhance student engagement with STEM learning areas and encourages students' view of themselves as competent learners in STEAM subjects. Generating knowledge through high-quality integrated STEAM learning does not necessarily lead to lowered academic standards and a watering down of learning objectives. Rather, informed STEAM planning seeks to address specific aspects of the curriculum and reflects developmentally appropriate teaching strategies for young children.

Learning and teaching are influenced by an evolving social and policy environment. This approach offers teachers an opportunity to reflect on their teaching practice. Reflective practice includes action: either to continue with existing practices or to change practice in light of new theories and research. As part of this process, teachers may reconsider what authentic learning looks like for young children and how this can be supported with developmentally appropriate pedagogy.

To be able to apply an Arts Immersion approach, teachers need to be confident in critically reflecting on their own practice and in collaborating with a more experienced teacher. It takes confidence for a teacher to avoid the dangers of adopting a defensive stance, a non-questioning attitude and resistance towards new ideas and emerging research. Rather, teachers need to be comfortable with allowing their students to become active learners through exploration and inquiry.

It is not necessary for teachers engaging in this type of professional learning to be experienced in arts education. However, they need to be open to engaging with other learning processes, skills and ways of knowing. In applying an arts Immersion approach, the role of a teacher includes helping students find the best answer rather than the quickest answer, providing a range of media with which students can communicate, and encouraging students to accept responsibility for their own learning. Teacher reflective practice is also called for. In enacting an Arts Immersion approach, teachers value a wide range of student responses to tasks, are comfortable with ambiguity, acknowledge and reflect upon their own areas of bias and celebrate individual and group cultural diversity by creating inclusive learning environments.

This arts-based approach to learning and teaching relies on a constructivist pedagogy in which learners *construct* knowledge through a series of experiences as they explore problems, ideas and issues. This contrasts markedly with a focus on having students memorize facts that have been provided by the teacher. Through authentic experiences, young children can take responsibility for their own learning by thinking deeply.

Engaging with concrete objects, acquiring knowledge and skills, communicating understanding and collaborating with others reflect active and immersive learning. Some teachers may view the body as subordinate to the mind and a playful distraction best left to young children.

However, teachers may benefit from acknowledging what early learners seem to know instinctively—that the world is first experienced in the body before it is processed in the mind. The cohesive use of the mind and body that characterizes embodied learning has benefits far beyond early childhood education.

Key findings from the case study:

- Initial introduction of Arts Immersion activities requires careful planning and delivery.
- Introductory activities should be simple and rely on existing resources.
- Beginning Arts Immersion teachers should be encouraged to share the responsibility of sourcing and designing activities with more experienced Arts Immersion teachers.
- Positive experiences with simpler Arts immersion strategies can pave the way to building confidence and competence.
- More complex Arts Immersion approaches can be developed over time with the assistance of experienced Arts Immersion teachers if required.

Suggestions for Family-Based Arts Immersion Activities

Learning is naturally integrated into everyday activities when family members explore concepts with children. Home-school partnerships are important for children's learning and the home environment offers endless opportunities for families to support learning, fostering a purposeful connection between mind and body through arts activities (such as painting, drawing, music, dance and drama). When children's learning is supported at home, children realize that learning does not only happen at school.

Some suggestions of Arts Immersed learning opportunities for young children are provided below.

- Establish a neighbourhood cooperative with a few families to learn about animals in the backyard or near the family home. Design, create and refine objects to support the welfare of these animals (e.g. create water containers, bird baths, feeders and safe spaces for animals).
- Create a book about a family holiday using text, photographs and drawings.
- Create family/community dances and movement activities to explain scientific concepts such as melting ice and water vapour. Incorporate music. These can also be recorded as a media arts artefact or presented at a family concert.
- Express a biological cycle as a series of family friezes/tableaux (drama). For example, show the stages of change in the life cycle of a frog (eggs, different stages of a tadpole, frog) or butterfly (eggs, caterpillar, chrysalis, butterfly). You could photograph the family frieze and display it at home or at your local library.
- Create an innovative book club with family and friends. At book club meetings, stories (or parts of stories) could be acted out, or people could take on a character from a book and be interviewed. Extend this idea by wearing costumes! Invent (or act out) different endings or change the characters (e.g. a big, bad wolf who is timid and likes to collect flowers).
- Invite local musicians to help with designing and creating musical instruments together (e.g. shakers or drums using different quantities of water in bottles to make different pitches, wind chimes, 'string' instruments with boxes and rubber bands or percussion instruments like drums and thongaphones).
- Use digital arts in the form of animation software to explain scientific, mathematical or engineering processes—for example, Renderforest, Biteable, Moovly, Animaker, Animatron, Toonator, Powtoon and GoAnimate. You could share these with experts in the relevant fields, in an online forum, or on an educational resource site.

- Play charades. Science-, technology-, engineering- and mathematics-related concepts and processes can be acted out using the language of dance, drama and music.
- Invent your own version of Pictionary. Science-, technology-, engineering- and mathematics-related concepts and processes can be drawn for people to guess. In this way, you incorporate visual arts with STEM.

Conclusion

This chapter has articulated the why, what and how of authentic STEAM approaches with children in the early years. The case studies presented demonstrate how a STEAM approach, as part of a broader Arts Immersion approach to learning and teaching, enhanced student engagement and improved outcomes in STEM subjects by reframing them as exciting and dynamic learning areas. Toddlers benefited from the purposeful inclusion of planning, refining, constructing and decorating bird feeders, using an interest in birds as the springboard for exploring engineering and technology concepts within an Arts Immersion approach. Primary school students were provided with an opportunity to learn by doing, and to represent their knowledge with words and pictures.

A STEAM-based approach opens up the possibility of multimodal and interdisciplinary learning and lends itself to teaching and learning across the early years from birth to age eight. It embraces student diversity, equity in the classroom, and creative problem solving with and through the arts.

References

Ang, L. (2014). Preschool or prep school? Rethinking the role of early years education. *Contemporary Issues in Early Childhood, 15*(2), 185–199. https://doi.org/10.2304/ciec.2014.15.2.185

Australian Curriculum Assessment and Reporting Authority. (2015). *Australian curriculum: The arts foundation to year 10—version 8.3*. Retrieved from https://www.australiancurriculum.edu.au/f-10-curriculum/the-arts/

References Withheld

Chapman, S. N. (2015). Arts immersion: Using the arts as a language across the primary school Curriculum. *Australian Journal of Teacher Education, 40*(9), 86–101. http://search.proquest.com/docview/1871585919/

Chapman, S. N. (2018). *Arts Immersion across the curriculum: An action research Case study in using the Arts as the home language in a primary school classroom.* Doctoral Dissertation, Griffith University, Brisbane, Australia. http://hdl.handle.net/10072/384903

Chapman, S. N. (2020). STEAMing ahead: Using the arts languages to engage students and enhance learning. *Practical Literacy: Early and Primary Years. 25*(2), 17–19.

Daugherty, M. (2013). The prospect of an "a" in STEM education. *Journal of STEM Education, 14*(2), 10–15. http://www.uastem.com/wp-content/uploads/2012/08/The-Prospect-of-an-A-in-STEM-Education.pdf

Department of Education Employment and Workplace Relations. (2009). *Belonging, being and becoming: The early years learning framework for Australia.* Canberra, ACT, Australia: Council of Australian Governments.

Doverborg, E., & Pramling Samuelsson, I. (2000). *Att förstå barns tankar. Metodik för barnintervjuer* (2:a rev. uppl.). Stockholm, Sweden: Liber.

Ewing, R., Miller, C., & Locke, T. (2014). Editorial: An Arts-led English and literacy curriculum. *English Teaching: Practice and Critique, 13*(2), 1–4.

Eisner, E. (2003). The arts and the creation of mind. *Language Arts, 80*(5), 340–344.

Eisner, E. (2005). Opening a shuttered window: An introduction to a special section on the arts and the intellect. *The Phi Delta Kappan, 87*(1), 8–10. https://doi.org/10.1177/003172170508700104

Garvis, S. (2012). Beginning generalist teacher self-efficacy for music compared with maths and English. *British Journal of Music Education, 30*(1), 85–101. https://doi.org/ 10.1017/S0265051712000411

Garvis, S., & Lemon, N. (2013). Are the arts important in schooling? Clear messages from the voices of pre-service generalist teachers in Australia. *Australian Journal of Music Education, 2*, 98–104.

Goodfellow, J. (2009). *The early years learning framework: Getting started.* Deakin West, ACT, Australia: Early Childhood Australia.

Guyotte, K., Sochacka, N., Costantino, T., Kellam, N., & Walther, J. (2015). Collaborative creativity in STEAM: Narratives of art education students' experiences in transdisciplinary spaces. *International Journal of Education & the Arts, 16*(15), 1–39. http://search.proquest.com/docview/1773231622/

Humes, W. (2013). Curriculum for excellence and interdisciplinary learning. *Scottish Educational Review, 45*(1), 82–93.

Kemmis, S., McTaggart, R., & Nixon, R. (2014). *The action research planner: Doing critical participatory action research.* Singapore: Springer. https://doi.org/10.1007/978-9814560-67-2

Klenowski, V. (2011). Assessment for learning in the accountability era: Queensland, Australia. *Studies in Educational Evaluation, 37*(1), 78–83. https://doi.org/10.1016/j.stueduc.2011.03.003

Pramling Samuelsson, I., Asplund Carlsson, M., Olsson, B., Pramling, N., & Wallerstedt, C. (2009). The art of teaching children the arts: Music, dance and poetry with children aged 2-8 years old. *International Journal of Early Years Education, 17*(2), 119–135.

Pramling Samuelsson, I., & Asplund Carlsson, M. (2008). The playing learning child: Towards a pedagogy of early childhood. *Scandinavian Journal of Educational Research, 52*(6), 623–641.

Schiller, W., & Meiners, J. (2012). Dance: Moving beyond steps to ideas. In S. Wright (Ed.), *Children, meaning-making and the arts* (2nd ed.) (pp. 1–29). Frenchs Forest, NSW: Pearson. https://doi.org/10.4225/35/5a39a42c85b42

Stolz, S. A. (2015). Embodied learning, educational philosophy and theory. *Educational Philosophy and Theory, 47*(5), 474–487. https://doi.org/10.1080/00131857.2013.879694

Taylor, P. (2016, August 7–9). *Why is a STEAM curriculum perspective crucial to the 21st century?* [Conference paper]. Brisbane, Queensland: Australian Council for Educational Research (ACER) Research Conference. https://research.acer.edu.au/cgi/viewcontent.cgi?article=1299&context=research_conference

Thompson, G. (2013). NAPLAN, MySchool and accountability: Teacher perceptions of the effects of high stakes testing in Australia. *International Education Journal, 12*(2), 62–84. http://search.proquest.com/docview/1651841160

Watson, A., & Watson, G. (2013, October, 1–4). Transitioning STEM to STEAM: Reformation of engineering education. *The Journal for Quality and Participation.* https://www.academia.edu/8766909/Transitioning_STEM_to_STEAM_Reformation_of_Engineering_Education

Wright, S. (2012). *Children, meaning-making and the arts* (2nd ed.). Frenchs Forest, NSW: Pearson.

9

Using Mathematical Investigations in Projects for STEAM Integration

Marianne Knaus

Mathematics as an Act of Inquiry

Mathematics is a way of describing the world we live in and is more about processes, structure, order and relationships rather than simply providing answers. In daily life, we can explore mathematics concepts all around us; maths is everywhere if you take the time to notice it (Knaus, 2013). From birth, babies are developing an awareness of mathematics, recognising shape, pattern and number. Children are busily engaged in exploring and finding out and not cognizant that they might be learning mathematics. In the early years, children do not necessarily recognise or compartmentalise subject areas such as maths, science, arts and technology. The focus is on the learning and inquiry to find out and know more about their world. It is parents and teachers who are able to elaborate on the child's inquiry to extend the learning and introduce mathematics concepts. When folding the washing,

M. Knaus (✉)
Edith Cowan University, Joondalup, WA, Australia
e-mail: m.knaus@ecu.edu.au

C. Cohrssen, S. Garvis (eds.), *Embedding STEAM in Early Childhood Education and Care*, https://doi.org/10.1007/978-3-030-65624-9_9

for example, a child may notice that there are two socks in each pile as a parent folds them into one bundle. The socks become groups of two. Experiences like this can start a conversation about counting in twos, or form the early basis of grouping and multiplication. Teachers and parents can make the most of incidental opportunities that occur every day.

Ruben's Encounter with a Puddle

On a walk with his parents, Ruben (16 months of age) noticed a puddle. Lately, the family has been on many walks and Ruben has been testing out his new gumboots in all the puddles he could find after the recent rain. He walks into the puddle immersing his boots right in the middle of the water, exploring the depth. He uses his hands to dig down deeper to find the bottom of the puddle (Fig. 9.1). He explores the leaves and grass, pulling them from the water. He then watches the water as it slowly drips off his fingers, making concentric circle patterns as each drop lands on the surface.

It is during informal, everyday opportunities like the above example, that mathematics can be noticed and discussed. The Let's Count programme in Australia successfully uses the mantra "notice, explore and talk about" in their workshops that bring together early childhood educators, parents and other family members. This programme has proven successful in enhancing children's mathematical engagement, learning outcomes and dispositions (Perry, Gervasoni, & Dockett, 2012). The term *mathematising* is used by Rosales (2015) to describe the way in which mathematics concepts are extended in contextual situations during opportune moments to help children understand the mathematics that is occurring. It is in real-world contexts that mathematics becomes meaningful and purposeful, and adults play an essential role in mathematising everyday experiences. However, in order to make the most of these learning experiences, it is necessary to have an understanding of mathematics concepts to be able to promote this learning.

The basic concepts children learn include aspects of number, geometry, measurement, algebra, data and probability. Experiences with these concepts will, however, look different for children aged under five years when compared with children at school. The way children start to make sense of the world and gain knowledge of mathematics is through a range of high-quality play opportunities. It is during play and everyday

Fig. 9.1 Ruben investigates a puddle

opportunities that children are purposefully engaged in problem-solving, discovering, reasoning, and using trial and error in ways that enable mathematics concepts to be explored. These concrete experiences are the building blocks of knowledge that are later transformed into more complex, abstract concepts. The path to concept development moves from informal understanding to formal knowledge according to Bruner's (1966) conception of the enactive-iconic-symbolic representation. A concept is first experienced through a physical experience; an *enactive* form. The enactive phase occurs between birth and one year of age. A typical example for babies would be when they shake a rattle. This is a physical action the baby has stored in their memory. The next step in the progression from concrete to abstract concepts is *iconic* representation, when an image is recalled. For example, a toddler may see a picture of a rattle in a book and point to it as it as they recall a rattle. The iconic phase uses visual images to assist the thinking process and is typically observed when a child is between one and six years of age. *Symbolic* representation, when a rattle is represented as a spoken word or written in a symbolic form, is observed at about six to seven years of age. When children begin to learn about mathematics, we provide concrete materials for them to explore, and with time and rehearsal opportunities, combined with the language of mathematics, more complex concepts can be developed. It is using the words alongside the 'doing' that is most important in concept formation.

The foundation for learning these concepts is the processes or skills which include identifying and describing attributes, matching, sorting, comparing and ordering (Irons, 2007). Attributes are particular characteristics of an object such as colour, shape or size. When playing and talking with babies, we encourage them to observe the attributes of objects: the ball is round, it is red. These conversations help them to distinguish attributes of objects by referring to similarities and differences. Matching, for toddlers and pre-schoolers, is finding two attributes that are the same. Matching is a necessary skill for learning one-to-one correspondence, and there are many ways we can encourage this skill through matching everyday items. Sorting also involves matching, but we use a greater number of objects. We can sort shells, buttons or leaves. When introducing sorting, we start off with one attribute at a time. Sorting becomes more complex when children learn to sort by two attributes. The process of comparing is important too: looking for differences between two or more items. Babies and toddlers are able to compare size differences such as big and small. With measurement, we compare length, weight, capacity, area and time, and pre-school children start to learn more complex measurement language. In number, children also compare sets or groups of objects. Ordering involves comparing and placing items into sequence, for example, from the shortest to the tallest. There are many items that can be ordered, from blocks and crayons to rocks and leaves. Groups of objects can be ordered too: a plate with one grape, a plate with two grapes and a plate with three grapes. The beginning processes support the development of mathematical ideas and more complex concepts across the year levels moving towards school entry. A breakdown of the concepts relative to age groups is shown in Table 9.1. Play alone does not ensure that children are learning mathematics. Parents, teachers and other adults can enhance children's learning by asking questions and mathematising their ideas (Table 9.1).

Table 9.1 Mathematics concepts relative to year levels

Concepts	Birth to 2 years	2–3 years	3–4 years	4–5 years
Number	Chants some counting words Differentiates quantities from 0 to 3 Compares small sets of objects Developing an understanding of "oneness", "twoness" Subitises up to 3 objects Recognises order in sets of size	Recites counting words (not necessarily in right order) Uses terms: more, same, different Subitises up to 4 objects Becomes more accurate in counting small numbers of objects Constructs equivalent sets of objects Adds one more	Recites counting words to 5, then 10 Uses number names in stable order Recognises numerals to 5 and that they represent numbers Subitises, names and selects 3 or 4 objects Partitions small numbers Draws symbols to represent numbers Recognises patterns on a dice Compares collections of objects and describes more or less	Counts beyond 10 Counts items using 1:1 correspondence Subitises to 5 Recognises numerals to 10 Represents cardinal number Counts on from small numbers Counts back from 10 Recognises decades 10, 20, 30, 40, 50… Skip counts in 2s, 5s, 10s Recognises patterns to 10 Partitions sets Combines two groups of objects to add or take away Shares objects equally into fair shares Divides sets into groups
Geometry (shapes)	Distinguishes between open and closed shapes Stacks 3D shapes Makes arrangements of shapes and lines Matches and names basic shapes (square, triangle)	Distinguishes between circles, squares and triangles Compares two shapes if they are same or different Stacks 3D shapes and builds towers and walls Matches two identical 3D objects Manipulates shapes and takes them apart	Recognises common shapes triangle, circle, square Explores and sorts shapes Constructs and deconstructs shapes Fits shapes together and takes them apart Matches shapes in puzzles and posting boxes	Recognises more shapes: hexagon, rhombus, trapezium Draws common shapes: triangle, circle, square Compares and classifies shapes Knows properties of simple 2D and 3D shapes (number of sides) Creates complex block play enclosures, arches, corners

(*continued*)

Table 9.1 (continued)

Concepts	Birth to 2 years	2–3 years	3–4 years	4–5 years
Geometry (spatial awareness)	Starts to get a sense of their own bodies and the relationship to the space around them Recognises patterns of movement and tracks objects with eyes Recognises and reaches for things in the distance Develops path integration Explores the characteristics of blocks Understands position words: on, in, under, up, down	Locates familiar objects in environment Points to familiar landmarks along a route Remember short journey sequences Stacks blocks vertically or horizontally in rows Creates enclosures with blocks Understands position words: beside, between	Locates objects when they have been moved Follows verbal directions Follows simple maps Bridging in block play to create more complex structures Builds models Understands position words: in front, behind	Gives and follows directions Locates items on a map Draws simple maps of routes Block play becomes more complex Rotates and recognises a shape Explores reflection and symmetry Rotates, flips, turns and slides puzzle pieces and shapes Understands position words: left and right

(continued)

Table 9.1 (continued)

Concepts	Birth to 2 years	2–3 years	3–4 years	4–5 years
Measurement	Explores attributes through sensory-motor play: reaching, moving, lifting, turning, covering, fill, empty	Recognises and names attributes: big, little, small, long, tall, high Becomes familiar with routine times	Compares items that are longer and heavier or the same Compares capacity pouring into containers Explores angles through block building Recognises time as a sequence of events	Makes estimations and comparisons of length, weight, capacity Makes ordered arrangements shortest to tallest using seriation Familiar with daily routine and days of the week
Algebra	Notices items with a fixed order Notices patterns in fabric and environment Recognises pattern and in familiar songs and rhymes Makes repeated patterns by shaking a rattle	Engages in repetitive tasks filling, emptying Lines up toys in repeating patterns Makes repeated marks in painting and drawing Creates imprints in dough	Verbalises before, after, next to, start, finish Makes line patterns Recognises a symmetrical pattern Copies and creates simple two-part patterns	Sorts and classifies familiar objects and explains the basis for these classifications Copies and extends a sequence Recognises a cyclical pattern Copies, continues, creates patterns with objects and drawings Recognises a growing pattern

(*continued*)

Table 9.1 (continued)

Concepts	Birth to 2 years	2–3 years	3–4 years	4–5 years
Data and probability	Discovers attributes of objects and names them	Recognises attributes such as colour, shape, size Compares objects that are same or different Matches two or more objects Makes purposeful collections Moves items from one group to another to sort them	Sorts, classifies and matches objects according to attributes Orders according to one attribute Groups items such as favourite pets	Uses the language of chance, possible, impossible, likely, unlikely Answers simple questions to collect information (yes/no) Represents data using concrete materials and interprets data Uses pictographic materials and interprets data

An Example of Mathematisation

"SHE'S 1, 2, 3, 4, 5, 6, 7, 8, 9, 10, 11, 12, 21...SHE'S 21 BIG."

The educators overheard an interesting dialogue taking place: a group of children were gathered around their whole-body portraits and were questioning the portraits' identities and genders.

Jamie: "How big is she?"
Alex: "She's bigger than me."
Riley: "She's taller than me, she's big!"
Asha: "Wait, stand back, I'll do it properly. She's 1, 2, 3, 4, 5, 6, 7, 8, 9, 10, 11, 12, 21...She's 21 big."

The children's interest then shifted to "how big" the whole-body was. The children were standing next to or in front of one whole-body portrait, comparing themselves and others to its size. The children used their hands and bodies to investigate "how big" the figure was; they did this both individually and collaboratively. In response to the children's evolving ideas, the kindergarten educators saw this as an opportunity to scaffold the children's learning by challenging their thinking and guiding their learning. They provided the children with investigation baskets containing multiple measuring tools, clipboards, paper and pencils.

When the children discovered the baskets of measuring tools, they immediately unrolled the measuring tapes and held them up to the whole-body portraits. The children shared their opinions on the length of the figures.

Asha: "Well, she is 60 long."
Riley: "Six – zero, that means sixty."

The children then took out the clipboards and documented their findings: Asha and Riley both drew an approximation of the number 60 (Acknowledgement: Nido Early School QV1).

Knowledge of mathematics concepts is a starting point for being able to mathematise. Rosales (2015) has developed a framework for mathematising as a learning process. The framework includes four components:

1. Observing—identifying and recognising a learning opportunity from which to create engaging mathematical learning experiences.
2. Exploration—promoting curiosity and learning with the use of materials to create, construct and develop thinking skills.
3. Language modelling—using appropriate language to label, question, expand and extend.
4. Inquiry—using strategies that lead to projects that incorporate higher-level thinking skills: problem-solving, reasoning, critical thinking and reflection.

Rosales' (2015) framework can be applied any time, anywhere, in an informal or a formal learning environment. Everyday routines and situations are ideal starting points for specific concepts to be discussed and mathematised, including with infants and toddlers who are investigating their world. The fourth component, inquiry, is explained in more depth below.

Inquiry Approaches

Mathematics learning through inquiry assists children to make meaning from their investigations within real-world contexts. An inquiry focuses on the learner as an active participant in knowledge acquisition, based on constructivist theories of learning advanced by Dewey (1938), Piaget (1952), Bruner (1966) and Vygotsky (1978). Using an inquiry approach has many benefits, and according to Rosicka (2016, p. 8), "Inquiry-based learning builds from a natural process of inquiry in which students experience a 'need to know' that motivates and deepens learning". The Organisation for Economic Co-operation Development (OECD) (2012) endorses inquiry-based learning as key to the development of twenty-first century skills which include critical thinking, creativity, collaboration and communication. Inquiry and exploration are the foundations of scientific learning and are essential features of STEAM. An inquiry approach encourages positive dispositions for learning and engagement in deeper levels of thinking that include predicting, critical thinking, developing hypotheses, reasoning and problem-solving. Emphasis is placed on the process skills of investigation, exploration and experimentation, as well as communication and sharing ideas collaboratively with others.

Babies and toddlers are engaged in inquiry as they explore and navigate their world. Parents and teachers can support this inquiry by providing an active learning environment with interesting materials, sensory opportunities and provocations (DECS, 2008). Interactions with babies and toddlers while engaged in exploration can motivate and inspire further curiosity, especially when adults mathematise child behaviours by making links to specific concepts (Knaus, 2013). Inquiries become more complex with pre-school children.

Building connections with families to learn about cultural practices, routines and children's interests is necessary to support mathematics inquiry. The contribution of the life experiences of children and families, and their contribution to learning, has been theorised. Funds of Knowledge is an educational theory that recognises the life experiences of children and families as a source of knowledge that provides many possibilities for positive pedagogical practices (González, Moll, & Amanti, 2005). The research of Hedges and Cooper (2014) further illustrates the influence of home learning environments on the interests and inquiries of babies and toddlers and consolidates this using the theory of funds of knowledge. Teachers are equipped to respond to children's interests and inquiries using authentic questions and meaningful learning when engaged in collaboration with families.

There are many different approaches and cycles that can be adopted to enact inquiry learning. The 5E Model was developed in 1987 by the Biological Sciences Curriculum Study (BSCS) and uses the structure of Engage, Explore, Explain, Elaborate and Evaluate (Bybee, 2014). Another inquiry approach is one developed by Pedaste et al. (2015) as illustrated in Fig. 9.2. In the first phase, the inquiry begins with *Orientation* during which the topic is introduced through a provocation, an inquiry question, or a problem. Phase Two includes the *Conceptualisation* of the investigation. This phase is based on questioning, with the assistance of a teacher, to guide the inquiry and generate a hypothesis. Phase Three is about *Investigation* and could take days, weeks or months depending on the topic, interest and age of the children. During this phase, teachers and children explore, experiment and collect data to try out their ideas and test their current understandings. It can also involve fieldwork, drawings, videos, photographs, documentation and visits from experts on the topic. Phase Four is the *Conclusion* to the inquiry at which point explanations are provided and an evaluation of the hypothesis undertaken.

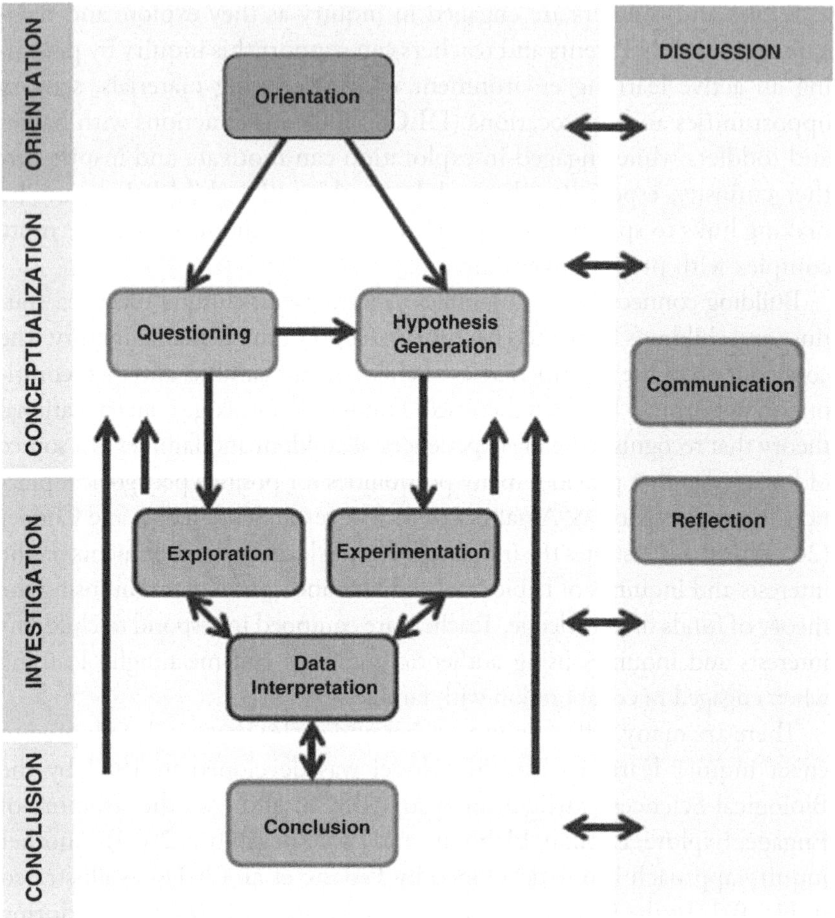

Fig. 9.2 Inquiry-based learning framework (Pedaste et al. 2015)

Discussion is incorporated in all of the phases of the inquiry. Here, communication and reflection with peers and teachers assist in explanation of their understanding through reflection and the sharing of ideas.

Kath Murdoch (2015) proposes an inquiry model with a sequence of phases (see Fig. 9.3) described as a journey of learning that can be conducted over a series of weeks. However, the Kath Murdoch approach is more than just going through the motions of a cyclical model; it positions understanding as the focus of the inquiry to help children inquire,

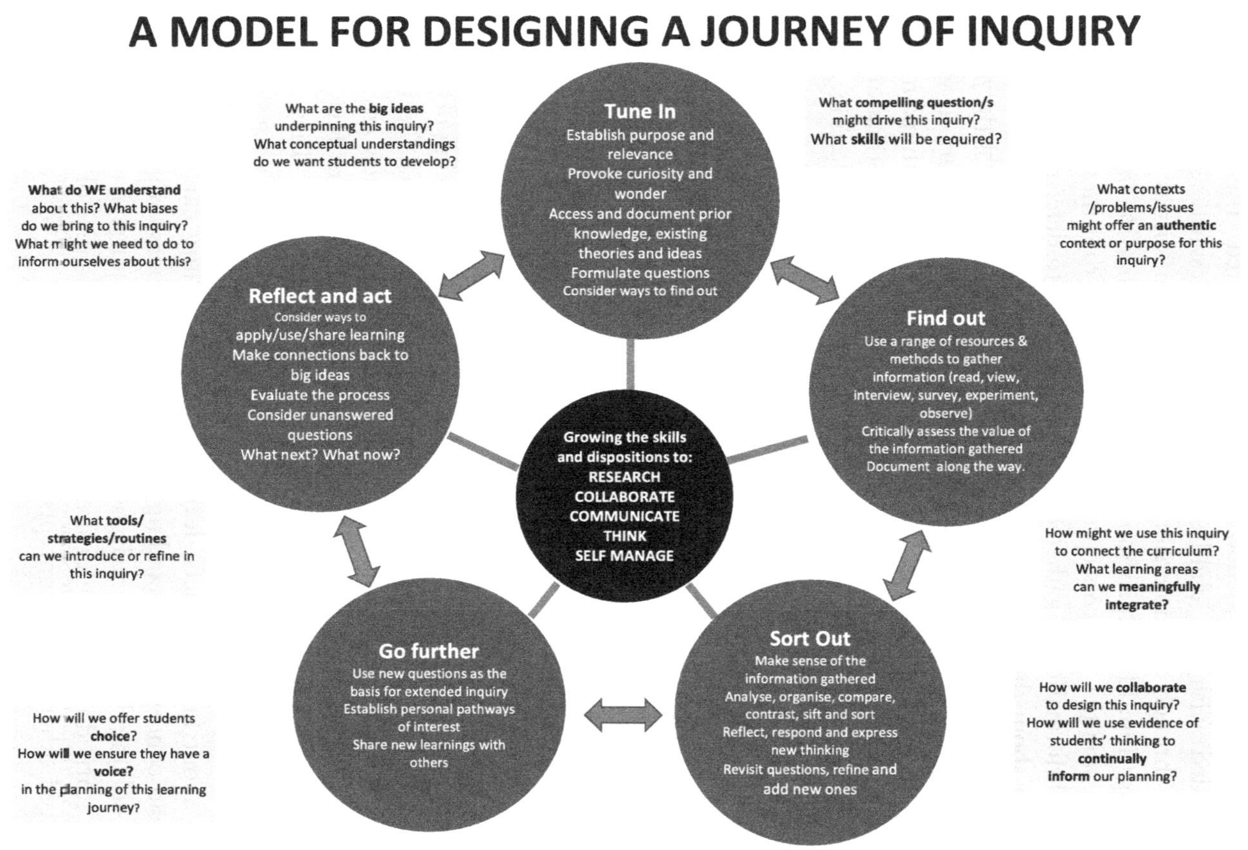

Fig. 9.3 A model for designing a journey of inquiry (Murdoch, 2019)

question and reflect as new learning is constructed. The planning is scaffolded with the children, but is not necessarily linear as it occurs through an immersion of phases and intentions as a process of investigation. Ideas, questions and suggestions *frame the inquiry* forming possibilities and conceptual underpinnings. From these initial big ideas, information is gathered with the children to determine what is known about the topic. Children and teachers *tune in* and make connections with previous knowledge and understanding. Next is the *finding out* phase with the gathering of new information in various ways and the recording of this information to help children think more deeply on the topic. Children then make sense of the information in the *sorting out* phase through analysing, comparing, reflecting and revisiting new thinking and previous questions. In the *going further* phase, children are encouraged to work independently to extend the inquiry. Children in the *reflecting and acting* phase review what they have learned and share this new knowledge, often in a collaborative way. Finally, *evaluating* is conducted by the teacher to review the effectiveness of the inquiry, assess the children and look back on the understandings gained through the journey of the whole inquiry.

Restricting mathematics learning to formal experiences only, does not enhance a child's inner drive to acquire new knowledge. Children start learning from a desire to find out and know as a consequence of their curious nature and do not conceive of learning taking place in subject silos. Krogh and Morehouse (2014, p. 1) state that "learning through inquiry begins at birth as infants explore their new environment and the people in it. Children learn more in their first few years of life than they will in any other development phase. They learn through continual inquiry and observation using all their senses, coupled with encouragement and modeling of adults". Mathematics is not just about knowing facts; learning through inquiry can enrich and extend concept understanding and in particular, reasoning and problem-solving. Using the pedagogical practice of inquiry, mathematics is best learned in meaningful situations through guidance of more experienced others where concepts can be connected to real-world problems and situations.

Intentional Teaching

During inquiry, parents, teachers and adults lead and guide the learning through sensitive intervention in appropriate and opportune moments so that the mathematics is made visible and relevant. The Early Years Learning Framework for Australia (Department of Education, Employment, and Workplace Relations [DEEWR], 2009, p. 15) recommends that to teach intentionally, we need to be "deliberate, purposeful and thoughtful" and to use "strategies such as modelling and demonstrating, open questioning, speculating, explaining, engaging in shared thinking and problem solving to extend children's thinking and learning". Siraj-Blatchford (2009) refers to sustained shared thinking to describe two people–often a teacher or a parent and a child–co-constructing learning by sharing knowledge and understanding to support and extend learning. The interchange of ideas could include discussing, challenging thinking, questioning and modelling to scaffold the child's learning. Inquiry learning is integral to intentional teaching and through ongoing reflective practice, teachers become "in-tune" with children's thinking and equipped to adapt pedagogical practices and strategies to the context.

Intentional teaching also involves setting up physical spaces. Effective mathematics inquiry takes place within active play environments that foster curiosity, wonder, problem-solving, persistence and confidence. Provision of suitable spaces and resources, as well as time to explore, are important considerations. Provocations and interesting materials provide stimuli for inquiry and questions. Children are able to lead investigations when they are encouraged to explore and collaborate with others in unhurried environments. They need time for ideas to develop and to engage more deeply in their thinking. Intentional teaching can be facilitated by more experienced peers as well as teachers and parents. Indeed, a study on inquiry learning conduct by Wu and Lin (2016) found that communication in mathematics was important and that children's peers were the best learning partners, particularly with regard to sharing and expressing ideas.

The use of language plays a critical role in mathematics learning and is an essential component of intentional teaching. For babies and toddlers, we model 'labelling language' to introduce mathematics language. From labelling, we extend the words we use with very young children to expand vocabulary (Rosales, 2015). The use of children's literature, puppets, poems and rhymes based on mathematics themes are very helpful in introducing new language for all ages. For example, singing the song, Five Little Ducks, and using puppets to demonstrate the ducks going over the hill and far away, provides a both a mental image and the words to introduce the concept of subtraction. There are many words that are specific to mathematics that can be quite complex as well as words that have dual meanings like 'odd', 'mass' and 'mean' and these can be confusing. When mathematical language is used in meaningful contexts, it assists children to make the link between the spoken word and the meaning of the words. Concepts are learnt when experienced in everyday experiences, and if this happens often, children will learn the vocabulary and understand the concept. Taking part in discussions helps children to make sense of the mathematics they are learning and to clarify concepts. Language is necessary to foster higher-order thinking in mathematics: to reason, explain, justify and reflect (Riccomini, Smith, Hughes & Fries, 2015).

The effective use of questions encourages language production and is an important component of intentional teaching. Questions are necessary for an inquiry approach as they encourage children's thinking and nurture curiosity. Murdoch (2015) suggests that questions themselves can be the focus of an inquiry, but we also need to help children learn *how* to pose questions. The types of questions children ask may sustain an inquiry over days and weeks. As teachers and parents we can encourage children to:

1. Manipulate prior information;
2. State an idea in their own words;
3. Find a solution to a problem;
4. Observe and describe an event or an object;
5. Compare two or more objects;
6. Give examples;
7. Explain their thinking;
8. Apply ideas to new situations;
9. Compare and find relationships;
10. Make predictions or inferences; and
11. Make a judgement (MacDonald, 2015, p. 26).

Good inquiry questions will provoke thinking and further exploration in all the STEAM disciplines.

Integrated Learning

Active learning through inquiry promotes a variety of ideas and perspectives from different disciplines. STEAM incorporates the disciplines of Science, Technology, Engineering, Arts and Mathematics. There is debate as to how much of each discipline needs to be included to call it a STEAM project. Some people argue that incorporating any two of the disciplines is regarded as STEAM (Bybee, 2013, Moomaw, 2013, Sanders 2009) while others recommend that *all* components of STEAM need to be integrated into a project (Dugger, 2010, Pelesko, 2015). Dugger (2010, p. 3) defines mathematics as “the science of patterns and relationships” and goes on to conclude that this definition is itself a perfect description of STEAM as a whole. The National Council of Supervisors of Mathematics and the National Council of Teachers of Mathematics (ND) identify the importance of the STEAM disciplines, but also uphold the role of mathematics underpinning a firm foundation at the centre of any STEAM programme. Children require sound mathematics skills to tackle problems and endeavour to find solutions. Other benefits to an integrated STEAM curriculum include increased motivation, increased problem-solving skills and an understanding of how and what is being learnt (Rosicka, 2016).

Moomaw and Davis (2010) suggest that connecting topics through an integrated curriculum is recommended practice in early childhood education. As children play and investigate, there is a natural crossover between science, technology, engineering, arts and mathematics. Learning is not compartmentalised into curriculum areas but woven into discovery and investigation. Using children’s natural curiosity, the learning is meaningful and relevant, and strengthened by intrinsic motivation. STEAM is recognised to have many more benefits than just teaching the academic skills of the five disciplines. STEAM is credited with contributing to the acquisition of twenty-first century skills, sometimes referred to as ‘employability skills’ or ‘soft skills’, such as problem-solving, collaboration, creativity and innovation (Timms, Moyle, Weldon & Mitchell, 2018).

Projects

Inquiry-based learning provides opportunities for children to identify solutions to problems. These investigations can lead to extended, ongoing investigations or projects to develop knowledge, understanding and skills: "learning experiences that extend beyond one-off activities, that can be repeated or returned to, and that lend themselves to ongoing involvement, encourage deep learning. The ongoing nature of such experiences ensures that children's engagement with them becomes deeper and richer" (Touhill, 2012, p. 2). A project can take weeks or months and cultivate learning in creative and interesting ways. Teachers can integrate specific curriculum goals within such inquiry projects, providing a holistic approach to learning.

Children are actively engaged in multi-disciplinary learning and supported to represent their thinking through displays and presentations. In line with this, mathematics can be incorporated in projects in many ways (Table 9.2).

Case Study of Nido Early School QV1

At Nido Early School QV1, engaging in research with and alongside the children is an integral part of the curriculum. This example of an inquiry project originated from the unique location of the early learning centre in the upper plaza of the QV1, a 43-story building in the heart of the Perth central business district. One of the Nido team documented the inquiry, demonstrating how STEAM opportunities can evolve from the everyday experiences of children and staff.

> We wanted to view the city through the eyes of the children and understand what was of significance to them. Whilst we were working on creating on a sense of belonging within our school, we were also part of a larger QV1 complex: the heart of the CBD. We wondered how we could make the most of our unique location allowing our children to be valued and visible occupants and participants within.
>
> What were the identified places they believed connected them to our community: our city, and how did these places connect them together?

Table 9.2 Mathematics concepts and their representations in projects

Mathematics concept	Representation
Number	Counting items and labelling them
	Using numerals to label items
	Using number words
	Counting and comparing sets of items
	Using operations such as addition and subtraction
	Using symbols for addition, subtraction, multiplication and division
	Representing part/whole items
	Using money
Geometry	Two-dimensional shapes
	Three-dimensional shapes
	Maps
	Diagrams
	Dioramas
	Symmetry
Measurement	Length
	Weight
	Area
	Volume
	Capacity
	Timelines
Algebra	Patterns
	Ordering
	Sequences in flow charts
Data and probability	Sorting attributes into categories for graphs
	Estimation
	Prediction
	'Fair shares'
	Frequency data represented in bar graphs, charts, Venn diagrams

Excursions became an integral part of our curriculum as we explored the Perth CBD and exposed our children to a variety of experiences, places and people in our city. By September, we became QV1 locals as we participated in all that it had to offer. We joined in NAIDOC and Springfest celebrations; we viewed live performances whilst sipping on our babyccinos at the real Mary Street Bakery and continued to visit our community garden regularly to plan, plant and marvel at the garden's growth, saying hello to the chickens inhabiting this space.

Fig. 9.4 A map of the city developed with the children identifying where Nido Early School is located in the QVI building

In order to build a sense of connectedness to "the helicopter school" aka Nido Early School and to identify connections between our homes, Nido QV1 and other places significant to their city, we embarked on a mapping project.

We revisited the photos of our children's homes and mapped them out according to their location in relation to Nido and the CBD. We linked them via transport modes as many children catch trains and buses to get to the city from their homes. Our map illustrations included familiar locations such as their families' work buildings, the City of Perth Library which we have visited and other sights they would see on the way to and from Nido (Fig. 9.4).

We could see that our excursions played a crucial role within our project, inspiring our children's work back at the centre and steering the direction of this venture. Our children started to take pride in the map, initiating discussions around the map by sharing information and the locations of their homes and other places of significance to them. With the group expanding, some families began to share their home languages with us: teach us simple words spoken at home.

Aaron: "I speak Australia."
Mr Gee: "Yeah, your grandad lives on a farm, so you use a lot of farm style language, don't you?"
Aaron: "Yeah like tractors and sheep…my mum told me that she wroted that to you."

Adding cultural elements to our map meant our children were exposed to different family structures and cultural differences. Yet the visual aid of a map served as a great tool for our children to understand that despite the differences, we all belonged to the same city.

To give children further opportunities and new perspectives the city view, we took them to QV1 rooftop viewing deck which allowed the children to have a bird's eye view and see many of the sights of the CBD from high above.

Reflecting on this experience, the group talked about what they remembered seeing from the top of the tower. These memories were then represented using the artistic medium of clay. Utilising the art studio, we used a projector to revisit some of the images of the views on a large scale. The children responded with excitement and enthusiasm and continued their city inquiry. We have created a gallery of our children's city representations for them to view and revisit daily.

We feel that this project is only just starting to bloom, and we are excited about the journey ahead. Our intent going forward is to find a way for our children not only to become active participants within their community, but also to create an educating community: where the educational quality and the rights of children are at the forefront.

The mapping project enabled children to explore many mathematical skills and concepts including geometry (spatial awareness, two- and three-dimensional shape, maps, symmetry), measurement (length, height, area), algebra (patterns, ordering, sequences), number (counting, recognising numerals, money), data and probability (prediction, estimation, sorting). Using an inquiry approach created opportunities for exploring learning through investigation of real-world experiences and the integration of the arts, mathematics, science, technology and engineering.

Implementing STEAM Through Inquiry Approaches

Each of the STEAM disciplines involves problem-solving, creative thinking, reasoning, analysis and communication. Through play and investigation, children are involved in these processes as they acquire new knowledge. Even very young children can learn through STEAM experiences.

STEAM Tips for Babies and Toddlers

- Make the most of everyday experiences and routines such as cooking, folding the washing and setting the table. Provide interesting mobiles to look at and toys to explore during nappy changing.
- Use routines as opportunities to talk about STEAM and to ask questions that stimulate conversation: when washing hands, ask where the water comes from and where it goes. Mealtimes, nappy changes and driving in the car all provide opportunities to ask questions.
- Provide opportunities for sensory exploration. Encourage children to feel, see, hear and smell while playing with sand, water, mud, clay, playdough, bubbles, slime, shaving cream and goop.
- Encourage play and curiosity by offering toys that foster trial and error, cause and effect, repetition, discovery, problem-solving, counting, learning about attributes, sorting and matching, filling and emptying.
- Offer materials that lead to discovery such as boxes, blocks, cardboard tubes, different types and sizes of balls.
- Provide outdoor experiences in the natural world: playing in snow or splashing in rain puddles. Natural materials such as shells, rocks, leaves, bark, seed pods and clay encourage new insights and exploration.
- Allow children to use tools like magnifying glasses, tongs, tweezers, eye droppers, digital devices and ramps.
- Encourage children to use art materials to draw, paint, cut and create collages.
- Provide opportunities for role play with dress-ups, scarves, shoes, telephones, teapots and cups, dolls and strollers.
- Play music and encourage your child to dance and sing, use musical instruments and explore sounds and rhythms.

STEAM Tips for Three- to Eight-Year-Old Children

- Integrate learning areas and curriculum across the five STEAM disciplines for children to make connections to real and meaningful experiences.
- Talk about the weather, seasons, birdlife, animals or natural cycles during everyday, outdoor experiences. While shopping, question where the vegetables, fruit or milk come from, and talk about how butter and cheese are made. This could lead to further investigation.
- Provide tools such as pulleys, levers, ramps and wheels in playful contexts such as the sandpit to encourage exploration of cause and effect, problem-solving and reasoning.
- Go on excursions to explore and discover authentic real-world STEAM learning. Excursions do not need to be whole-class trips to a zoo; an excursion as simple as a walk to a park may also provide stimuli for new learning. As part of a project, excursions or incursions can orient children to first-hand observations and experiences. Invite experts into your centre to talk to the children. If a project is investigating how bread is made, visit a local bakery.
- Creative art experiences such as box constructions, printing, painting, drawing, sculpting can all lead to thinking across all STEAM disciplines.
- Dramatic play, music, dance, choreography encourage children to express their feelings, explore location and direction, and to innovate.
- Materials to design, construct and build such as blocks, manipulatives like LEGO and construction kits encourage engineering, mathematics and science.
- Hands-on opportunities to produce constructions by having a carpentry bench with timber, nails, hammers, screwdrivers, saw, a vice and recyclables will require close supervision but also encourage creativity and innovation.
- Set up a tinkering space for children to create and invent. A myriad of objects could be included such as magnets, circuits, robotic toys, LEDs, alligator clips, recyclable materials and items to join things together such as blue tac, playdough, and tape.

- Provide outdoor experiences in nature to explore shadows, wind, weather, reflections, or to observe plants, insects, birds and animals. Extend on these based on children's interests and motivation.
- Design activities that provide opportunities to measure, weigh, sequence, sort, order, classify and compare.
- Use children's literature to inspire STEAM opportunities, for example, *Who Sank the Boat*, *Mr Archimedes' Bath*, *There's a Hippopotamus Sitting on My Roof Eating Cake*, *What the Ladybird Heard* or *Alexander's Outing*.
- Provide provocations to invite learning and provoke thinking such as a light table with a collection of fascinating objects. The collection could be based on observations of children's interest such as a bird's nest, a bird's eggs, and related photographs. Provide clipboards for children to draw, write and ponder. A pushbike for children to dismantle and explore, or an engine to examine, invite many possibilities.

Any of the suggestions above could become the impetus for project work in which children take the lead in their learning, supported by teachers or parents. When STEAM experiences are purposefully planned, they provide meaningful opportunities for observing, making predictions, discussing, carrying out experiments, forming questions, finding patterns and building theories. Open-ended opportunities assist children to engage in planning, decision-making, and collaborating. All provide opportunities to embed mathematic thinking, making connections across the STEAM disciplines.

References

Bruner, J. S. (1966). *Toward a theory of instruction*. Cambridge, MA: Harvard University Press.

Bybee, R. W. (2013). *The case for STEM education: challenges and opportunities*. Arlington, VA: NSTA Press.

Bybee, R. W. (2014). *The BSCS 5E instructional model: Personal reflections and contemporary implications*. Science and Children, Apr.–May. http://go.galegroup.com/ps/i.do?id=GALE%7CA377575190&v=2.1&u=cowan&it=r&p=AONE&sw=w&asid=4c7b596fb31108ad66a2200428b98476

Department of Education and Children's Services South Australia. (2008). *Assessing for learning and development in the early years using observation scales: Reflect, respect, relate*. Hindmarsh, Australia: DECS Publishing.

Department of Education, Employment and Workplace Relations (DEEWR). (2009). *Belonging, being and becoming: The early years learning framework for Australia*. Canberra, ACT: Commonwealth of Australia.

Dewey, J. (1938). *Experience and education*. Springfield, OH: Collier Books.

Dugger, W. E., Jr. (2010). *Evolution of STEM in the United States*. Sixth Biennial International Conference on Technology Education Research.

González, N., Moll, L. C., & Amanti, C. (Eds.). (2005). *Funds of knowledge: Theorizing practices in households, communities and classrooms*. Mahwah, NJ: Lawrence Erlbaum.

Hedges, H., & Cooper, M. (2014). *Inquiring minds, meaningful responses: children's interests, inquiries, and working theories*. Wellington, New Zealand: TLRI.

Irons, R. (2007). *Mathematics for young minds: Beginning processes*. Brendale, Australia: Origo Education.

Knaus, M. (2013). *Maths is all around you: Developing mathematical concepts in the early years*. Albert Park, Australia: Teaching Solutions.

Knaus, M. (2017). Supporting early mathematics learning in early childhood settings. *Australasian Journal of Early Childhood, 42*(3), 4–13.

Krogh, S., & Morehouse, P. (2014). *The early childhood curriculum: Inquiry learning through integration* (2nd ed.). New York/London: Routledge.

MacDonald, A. (2015). *Investigating mathematics, science and technology in early childhood*. Oxford, UK: Oxford University Press.

Moomaw, S. (2013). *Teaching STEM the early years: activities for integrating science, technology, engineering, and mathematics*. St. Paul, MN: Redleaf Press.

Moomaw, S., & Davis, J. (2010). STEM comes to preschool. *Young Children, 65*(5), 12–18.

Murdoch, K. (2015). *The power of inquiry: Teaching and learning with curiosity, creativity and purpose in the contemporary classroom*. Northcote, Australia: Seastar Education.

Murdoch, K. (2019). *Updated diagram – Designing a journey of inquiry 2019*. https://www.kathmurdoch.com.au/

National Council of Supervisors of Mathematics and the National Council of Teachers of Mathematics. (n.d.). *Building STEM education on a sound mathematical foundation: A joint position statement on STEM from the national council of supervisors of mathematics and the national council of teachers mathematics*. https://www.nctm.org/uploadedFiles/Standards_

and_Positions/Position_Statements/Building%20STEM%20Education%20on%20a%20Sound%20Mathematical%20Foundation%20(NCSM-NCTM%202018).pdf

Organisation for Economic Co-operation Development (OECD). (2012). *The nature of learning: Using research to inspire practice: Practitioner guide from the innovative learning environments project*. Paris: OECD Publications.

Pedaste, M., Maeots, M., Siiman, L. A., de Jong, T., van Riesen, S. A. N., Kamp, E. T., et al. (2015). Phases of inquiry-based learning: Definitions and the inquiry cycle. *Educational Research Review, 14*, 47–61.

Pelesko, J. A. (2015). *STEM musings*. Model with mathematics website: http://modelwithmathematics.com/2015/11/stem-musings/

Perry, B., Gervasoni, A., & Dockett, S. (2012). Let's count: Evaluation of a pilot early mathematics program in low socioeconomic locations in Australia. In J. Dindyal, L. P. Cheng, & S. F. Ng (Eds.), *Mathematics education: Expanding horizons (Proceedings of the 35th annual conference of the Mathematics Education Research Group of Australasia)* (Vol. 2, pp. 594–601). Adelaide, Australia: Mathematics Education Research Group of Australasia.

Piaget, J. (1952). The origins of intelligence in children (8, 5, pp. 18–1952). New York: International Universities Press.

Riccomini, P. J., Smith, G. W., Hughes, E., & Fries, K. M. (2015). The language of mathematics: The importance of teaching and learning mathematical vocabulary. *Reading & Writing Quarterly, 31*(3), 235–252. https://doi.org/10.1080/10573569.2015.1030995

Rosales, A. (2015). *Mathematizing: An emergent math curriculum approach for young children*. St. Paul, MN: Redleaf Press.

Rosicka, C. (2016). *From concept to classroom: translating STEM education research into practice*. Camberwell, Australia: Australian Council of Educational Research (ACER).

Sanders, M. (2009). STEM, STEM Education, STEMmania. *The Technology Teacher, 68*(4), 20–26. http://hdl.handle.net/10919/51616

Siraj-Blatchord, I. (2009). Conceptualising progression in the pedagogy of play and sustained shared thinking in early childhood education: A Vygotskian perspective. *Educational and Child Psychology, 26*, 2.

Timms, M., Moyle, K., Weldon, P., & Mitchell, P. (2018). *Policy insights: Challenges in STEM learning in Australian schools*. Issue 7 May 2018. Camberwell, Australia: Australian Council for Educational Research.

Touhill, L. (2012). *Inquiry-based learning.* National quality standard professional learning program e-Newsletter no. 45. http://www.luketouhill.com.au/downloads/1791122/NQS_PLP_E-Newsletter_No45.pdf

Vygotsky, L. S. (1978). *Mind in society: The development of higher psychological processes*. Cambridge, MA: Harvard University Press.

Wu, S.-C., & Lin, F.-L. (2016). Inquiry-based mathematics curriculum design for young children: Teaching experiment and reflection. *Eurasia Journal of Mathematics, Science & Technology Education, 12*(4), 843–860.

10

Toddlers' Mathematics: Whole Body Learning

Karin Franzén

Introduction

Internationally, many studies in preschool mathematics have focused on older children. However, there is growing interest in the mathematical thinking of children aged under three years, as well as what this means for teaching practice in the context of play-based learning. This chapter addresses the M of Science, Technology, Engineering, Arts and Mathematics education (STEAM): mathematics. We will focus on the ways in which a toddler, Sara, used her body to develop mathematical knowledge. In the first example, Sara extracts a ball from under a cabinet. In the second example, she explores the size of a big wooden car. Both examples demonstrate an understanding of shape, location and direction, as well as problem-solving and perseverance. A child may not be aware that they are using mathematical thinking to solve a real-world problem. However, when teachers and parents recognise this behaviour, talk about

K. Franzén (✉)
Karlstads University, Karlstad, Sweden
e-mail: karin.franzen@kau.se

C. Cohrssen, S. Garvis (eds.), *Embedding STEAM in Early Childhood Education and Care*, https://doi.org/10.1007/978-3-030-65624-9_10

what the child is doing and praise perseverance, they encourage independent problem-solving, exploration, mathematical language acquisition and positive learning dispositions.

Let us set the scene for this chapter by meeting Sara. She is 13 months old and has recently started attending preschool in Sweden. Although she has been walking for a short time, Sara often crawls because it is a more efficient way for her to get around.

> *Sara crawls into the playroom. Suddenly, she stops. Sara has noticed a ball beneath a cabinet. She lies on her stomach and peers under the cabinet. Then, she bends her head to one side and tries to crawl beneath the cabinet. She bumps her head and stops moving. She lies still for a moment then she lowers her head closer to the floor, before wiggling further. She stretches out one arm as far as she can and pushes the ball with her fingertips. It rolls out from beneath the cabinet. Sara wiggles backwards, with her head tilted to one side. When she emerges from under the cabinet, she sits up, turns to her teacher and smiles.*

The Scandinavian countries have a strong social pedagogy tradition and a holistic approach to learning. Further, play and learning are regarded as inextricably linked. In line with the Swedish Education Act (2010.800) and the Swedish Curriculum for the Preschool (SNAE, 2018), early childhood education in Sweden is characterised by child-centred practice. The curriculum development goals are broad, and the child's own interest has great impact (Bennett, 2005; Broström, 2017). Thus, the curriculum emphasises the importance of following children's interests, and preschool teachers are required to integrate early mathematical education in children's play activities. It also requires teachers to understand what a child knows already and what they are ready to learn next. Evidence of a child's existing capabilities informs learning goals, and carefully planned play-based opportunities are the vehicles for learning (Anders & Rossbach, 2015; Björklund, 2013; Fisher, Hirsh-Pasek, & Golinkoff, 2012; McCray & Chen, 2012; Opperman, Anders, & Hachfeld, 2016).

Some early childhood teachers and parents may believe that supporting the mathematical thinking of children under three years is unnecessary. However, research shows a relationship between children's early mathematical knowledge and their later mathematical achievements (Bailey, Siegler, & Geary, 2014; Claessenn & Engel, 2013; Watts,

Duncan, Siegler, & Davis-Kean, 2014). Moreover, the period from one to three years of age may be a critical period for exploring mathematical concepts (Sheridan, Pramling Samuelsson, & Johansson, 2009). This is not to suggest that exploring mathematical concepts should be anything other than playful, as play-based learning is both developmentally appropriate and supports learning dispositions. Indeed, the Swedish Curriculum for the Preschool states:

> All children should experience the satisfaction and joy that come from making progress, overcoming difficulties and being an asset in the group. The preschool should give children the opportunity to develop a positive perception of themselves as learning, creative individuals. Children should therefore have the opportunity to discover and marvel, try and explore, and also to acquire and shape different skills and experiences. (SNAE, 2018, p. 10)

Let us return to Sara to see how her actions relate to this statement. She overcame the challenge of reaching the ball. Achieving this independently, both reflected and nurtured Sara's positive perception of her own ability to be creative in solving the problem. In future, Sara is likely to bend her head very low when crawling under furniture to avoid bumping her head and she will know that to reach a distant object, she needs to stretch her body as far as she can.

We could also use a mathematical lens in order to interpret Sara's actions. To do this, we need to know what mathematical thinking is associated with toddler development. However, it may be necessary for parents and early childhood teachers to reframe early mathematics learning (Nichols, Levay, O'Neil, & Volmert, 2019). Mathematical thinking does not start when children start school. Instead, it occurs along a continuum that starts at birth (Baroody, Clements, & Sarama, 2019). Clements and Sarama (2014) propose a learning trajectory approach that describes child behaviours (as indicators of emerging capabilities), aligning these with number knowledge; verbal and object counting; comparing, ordering and estimating; spatial thinking; knowledge about shapes labelling shapes as well as understanding what shapes are created when shapes are put together (or split into parts) and measurement (amongst other capabilities). In addition, they also discuss other mathematical processes and practices that children demonstrate and rehearse (again, along a continuum that starts at birth) such as reasoning, critical thinking and problem-solving.

In describing Care, Development and Learning, the Swedish Curriculum for the Preschool (SNAE, 2018) describes the following mathematics-related goals for children (pp. 14–15):

> The preschool should provide each child with the conditions to develop:
>
> - an ability to use mathematics to investigate, reflect on and try out different solutions to problems raised by themselves and others,
> - an understanding of space, time and form, and the basic properties of sets, patterns, quantities, order, numbers, measurement and change, and to reason mathematically about this,
> - an ability to discern, express, investigate and use mathematical concepts and their interrelationships.

What does Sara know? Rautio (2014) explains that adults "need to trust that the interaction between children and the world, seemingly irrational and mostly unreflecting, also has value" (p. 402). When we look closely at Sara's behaviours, we see that they are rational and goal-oriented. When analysing her efforts to reach the ball, a four-step process for problem-solving is observed: understanding the problem, devising a plan, carrying out the plan and looking back to check (Pólya, 2004). Sara *understands she has a problem* when she realises she cannot reach the ball. She knows the ball is *under* the cabinet and that the bottom of the cabinet is close to the floor. Sara *devises a plan* to reach the ball. This is evidenced by her lying on her stomach and turning her head to one side *before* reaching for the ball. Sara knows that the ball is far away from her. This is evidenced by her immediately stretching as far as she can and without grasping at the ball, pushing it with her fingertips, applying force to propel it out from under the cabinet. The process of *carrying out the problem*, when she moves the ball out from under the cabinet, was an opportunity for Sara to investigate and try out different solutions independently. The extent to which she reflected on them is beyond the scope of what may be inferred from this brief snapshot. However, given that Sara appeared to know immediately what to do suggests that she may have prior experience with similar problems to which she *could be looking back* since she has encountered similar challenges in the past. Sara demonstrates an understanding of space and form—she knew that the ball was beneath the cabinet, and she understood that balls roll when pushed.

Let us take these learning goals one step further to determine what the shape, spatial thinking and measurement developmental progression may look like for toddlers. Clements and Sarama (2014) suggest that by the age of two years, children typically recognise shapes that are the same or different. They are likely to be able to match shapes like squares, circles and triangles that are the same size and initially, the same orientation (e.g. both triangles are "standing" on a flat side). Children under the age of two years are unlikely to recognise length as an attribute of an object. Toddlers use a distance landmark to find an object and as they acquire the locational language of "on," "in," "under" and so forth, they learn how to use shapes or blocks to copy a picture or a simple construction. However, teachers may require support in recognising and responding to children's demonstrations of mathematical thinking. Here, the greater a preschool teacher's own mathematical concept knowledge, the better they are recognising mathematics in children's play (Opperman et al., 2016). Without this mathematics concept knowledge, a child's actions may be perceived to be "too mundane, too obvious, too pointless, or too insignificant to write about, explain, or even think about" (Horton & Kraftl, 2006, p. 71).

Of course, the reader recognises that Sara's actions were significant and so our attention now turns to the way in which toddlers use their bodies for learning in a way that adults have long since abandoned!

Toddlers' Learning

Sara lay on the floor beside the cabinet and turned her head. After bumping it, she turned her head more, wiggled forward, reached out her arm and pushed with her fingers. Her body was thus the instrument of her learning. Where adults and older children use their eyes to observe and comprehend attributes of objects, very young children use their bodies. When older children see a doll's chair, they know it will be too small to sit on; toddlers, on the other hand, need to try this out (Franzén, 2015a). A toddler's body is thus a tool for meaning-making.

Whilst we have reflected on typical learning trajectories and commented on opportunities for adults to recognise and support learning by narrating the child's actions using mathematical language, this approach is underpinned by the notion that learning is essentially an internal mental process that is influenced by prior experience and shaped by social interaction, which includes language modelling and language acquisition. However, learning is more than language and thinking (Franzén, 2014, 2015a, 2015b; Lenz Taguchi, 2010, 2011; Merrell, 2003; Palmer, 2011; Rautio, 2014; Yelland, 2010). Let us now consider an alternative view of learning.

Merrell (2003) asserts that nothing can be attributed only to the mind or to the body. Rather, there is reciprocity between the mind and the body that makes it impossible to separate what we learn with our bodies from what we learn by thinking. It can be understood as a reciprocal flow, a fusion of spirit and body (Merrell, 2003). Applying this theoretical approach to understand Sara's behaviours, we recognise that her understanding of the need to lower her head, lower it further and then to reach as far as she could to push the ball, reflects a reciprocal flow between her thinking and her physical experience. Macedonia uses the concept of "embodied cognition" and claims that the mind is integrated into the body's sensorimotor system (2019, p. 3):

> In fact, if we observe how children acquire language, they perform a multitude of sensorimotor acts. Children hear and repeat sequences of sounds (words), i.e., symbols but these symbols are related to objects they perceive with their senses or to actions they perform.

Young children touch, smell and mouth objects. Thus, a word symbolises a sensorimotor network that reflects all experiences, including the physical experiences encapsulated in the concept (Macedonia, 2019). For example, the concepts of low and high are bodily experiences as well as cognitive knowledge. Whether one prefers the notion of reciprocal flows or embodied cognition, common to both is the suggestion that a child's learning should offer opportunities for physical exploration of their worlds as learning results from thinking and physical experience. Certainly, as the child interacts with her environment, by thinking and by using her body, learning is prompted (Franzén, 2014, 2015a, 2015b), and in this way, the environment plays an important role in developing a child's knowledge. According to Barad (2007), meaning-making is influenced by objects, the environment, teachers, peers and the child's thoughts and feelings. Indeed, Barad argues that the child intra-acts (rather than interacts) with the environment since objects, the environment and the body *all* participate in the learning situation to affect meaning-making—learning.

Parents and early childhood teachers know that very young children use all available senses to experience their environments—this sensory exploration includes sight, hearing, smell, taste and touch. A child may not be able to describe an experience using spoken words but many researchers have explained how gestures help children and adults to communicate concepts—particularly spatial concepts (e.g. Ehrlich, Levine, & Goldin-Meadow, 2006; Elia, Gagatsis, & van den Heuvel-Panhuizen, 2014; Hedge & Cohrssen, 2019; Ontario Ministry of Education, 2014; Owens, 2015).

If we take an intra-active perspective, does Sara's learning become visible? Sara gained experience in the mathematical concepts of shape, location and direction. Using this perspective, we observe intra-action between Sara's understanding, Sara's body, the ball and the cabinet. From this perspective, it is not only the child's thinking and language that affect the learning situation but also objects in the environment, her peers, her teachers, the child's emotions and the preschool context, all contribute to the creation of knowledge.

In this way, even inanimate objects may be understood as active agents as they are co-producers in the production of meaning. This perspective may offer a better understanding of very young children's learning processes: learning is an intra-action between many different things, both human and non-human. If the ball had not been under the cabinet, Sara would not have had that opportunity to examine mathematic concepts in the way she did. Drawing on the notion of intra-action, the ball can be seen as having a form of agency since it participates in the learning experience. The ball lies there and Sara wants to play with it. In order to do so, Sara must work out where her body is in relation to the ball and how to change the position of her body in space to reach it. When she first tries to reach the ball, Sara's head bumps the edge of the cabinet—in this way, from an intra-active perspective, the cabinet contributes to her learning. Sara learns from this, recognising that the gap between the floor and the bottom of the cabinet is smaller than she had first understood and repositions her body in order to wiggle under the cabinet to reach the ball. She already knows about the attributes of the ball—that balls roll—and so she knows that if she pokes the ball a little it is likely to roll out from under the cabinet. Sara's mood and feelings also play a role in what she does and how the situation develops. Exercising emotional self-regulation and goal-directed behaviour throughout this process, she does not ask for help. Instead, Sara perseveres and consequently experiences success. This success is likely to be motivating and to reinforce her sense of agency and self-efficacy.

Observing Intra-active Learning

If toddlers' learning is intra-active, what does this mean in practice? Let us consider examples from one Swedish research project. Seven teachers working at four preschools participated in interviews that focused on the provision of mathematical learning experiences to children from 12 months to 3 years of age (Franzén, 2014). Teachers were asked to share how they include opportunities for mathematical learning in their planning, whilst following children's interests. During these interviews, teachers referred to the importance of supporting children's mathematical thinking, choosing developmentally appropriate and aesthetically pleasing resources for play. They spoke of the way in which the physical environment was set up to support learning and development. The teachers also emphasised the importance of modelling specific vocabulary when narrating their own, or the toddlers' behaviours. For example, they would say, "Now we will try to climb *up high*," and "We put your shoes *under* the shelf." They reported that hearing the word, whilst experiencing the concept, supported toddlers' learning. Several teachers commented on the crucial role *played by the toddlers' bodies* in learning about space, shapes, location and direction. Additional examples are provided below.

Spatial Location in Relation to Oneself

When looking at illustrations in a storybook with a toddler, telling the child that a car is parked in front of a house may not support the child's understanding of the spatial relationship between the car and the house as an illustration is not three dimensional. It is more meaningful to a child to explain the locational concept, "in front of" by referring to children's own bodies: "Look, Eric, you are standing *in front of* Hannah! Hannah, you are standing *behind* Eric." The younger a child is, the more the child learns with help of their body—as children mature, language alone may be enough to support understanding (Franzén, 2014).

Using the Body as a Unit of Measurement

Two children (both two years old) are playing with a box. They take everything out of the box and before both climbing into it, they jump and laugh, attracting their friends' attention without using words. However, sitting down in the box poses a problem because there is not enough room for their legs. At first, the children kick each other. After a while, they work out how to position all four legs inside the box. The problem is solved, and they laugh again. However, a third child starts to climb into the box. One of the children already inside the box shakes her head and says firmly, "No place, no place." Note that the child does not say, "No, no," but, "No place, no place." This suggests that having experienced the challenge of fitting two bodies (and four legs!) into the box, the children had investigated and learnt about the dimensions of the space within the box. Here, it is not suggested that formal measurement had occurred. Rather, the children have learnt that if they are to sit down inside the box, it is only big enough for two children. Here, the informal unit of measurement is the child's body. Further, this aligns with the progression point along the trajectory for volume for children aged under three years: the children have identified capacity of volume as an attribute of the box (Clements & Sarama, 2014, p. 207).

Let us return to 13-month-old Sara. She notices a large, wooden car that is popular with the children. On seeing the car, Sara drops to her hands and knees and crawls quickly towards the car. Crawling quickly (rather than slowly) indicates an understanding of time: reaching the car before other children will mean that she has the first turn. On reaching the car, she stands up on wobbly legs before trying to climb into the car. She makes a few attempts but does not lift her leg high enough. Toddlers often measure space with their bodies (Franzén, 2014, 2015a), and these repeated attempts to raise her leg up and over the side of the car are evidence of Sara using her body to understand height. Finally, Sara manages to get one foot into the car. Now she is standing with one foot inside the car and one outside. She stands still for a little while, then she leans her body forward, using it as a lever to swing her other leg into the car. Once inside the car, Sara sits in the driver's seat. She begins to "drive," turning the steering wheel and simultaneously following the direction of the steering wheel with her whole body. Suddenly, she stops driving. Something is troubling her. She turns and looks behind her, first to one side and then to the other. With much effort, she climbs out of the car again. She walks around the whole car, keeping one hand on the side of the car. In some places, she needs to bend down a little to follow the car's shape with her hand. After walking all around the car, Sara climbs into it again, sits down and begins to "drive" again. She is smiling and looks pleased with herself.

What was Sara doing? As very young children do not yet communicate their thoughts with spoken language, early childhood professionals and parents observe children and make inferences based on the ways in which children use their bodies for communicating and as tools for learning. Consequently, we infer that when Sara is inside the car, she is experiencing the car's size. She can see that the car is big when she is seated inside it, but she is not content with this. She needs to examine the size further. By walking around the car (bending down at times), and with help of her hand touching its surface, Sara understands more about its size and shape. (For additional analysis of Sara's intra-action with the wooden car, see Franzén, 2015a.)

Sensory Exploration of Shapes and Textures with Feet

Co-teachers in one early learning setting described toddlers playing with plastic construction bricks, taking off their socks to walk on the bricks with bare feet. Initially, the teachers were concerned that the children were too young to use the bricks and discussed removing them to protect children's health and safety. However, on reflection, the teachers agreed that whilst many children use their hands to examine objects, some use their mouths and others like to use their feet. The teachers decided to take a strength-based approach, left the bricks in the play area and ensured that close supervision was provided. This decision reflected their understanding that the sensory experience of walking on the bricks and feeling their differing sizes and shapes would be fundamentally different from the experience of handling the bricks with fingers and would thus support learning.

Visual Stimuli to Encourage Pattern-and-Structure Play

Teachers at one preschool reported that children showed no interest in exploring a large jar of coloured beads. Noticing this, the teachers sorted the beads by colour into two different jars. When the children saw pink beads in one jar and yellow in another, they began to use them. The children worked at patterning for extended periods of time, setting up opportunities for teachers to talk with the children about the structure of their patterns. Counting the beads also created opportunities for children to rehearse the counting principles (Gelman & Gallistel, 1978).

Thoughts and Suggestions for Parents and Teachers

One frequently hears that "mathematics is everywhere," but as teachers and parents, we need to *notice* toddlers' demonstrations of mathematical thinking in their spontaneous play in order to encourage this thinking and to model mathematical language. Some thoughts and suggestions for parents and teachers are provided below.

- Remember that toddlers use their bodies as tools for understanding mathematical concepts. Be alert to the problems that a child is trying to solve and use this as your starting point when modelling mathematical language and encouraging mathematical problem-solving. For example, you could say, "How many children are going to have apples? Apple children, let's stand here. Orange children, let's stand here. Are there more apple children or more orange children? How do we know? This group has more children. Let's count to check." Note that this approach takes time.

- Help toddlers to express their thinking, to persevere (with your support) and to problem-solve. This supports emerging language skills, sustained attention and positive attitude to learning. It also gives them agency over their own educational development.
- Talk to toddlers about what they are doing and what you are doing. Describe shapes and attributes of shapes. (For instance, "Your ball is round. It will roll. That block has flat sides. It doesn't roll.")
- Talk to toddlers about where things are (e.g. "Your teddy is sitting on the big, grey couch").
- Use number words in a practical context. (For instance, "Look at the big group of children! Let's count them. One, two, three, four, five. Five children!")
- Recognise that mathematics learning is more than numbers, counting and naming shapes. Learning occurs across different mathematical strands simultaneously. Children are learning about number and quantity at the same time as they are learning about shapes and measurement. They are also learning to find answers to simple questions (e.g. "Who has the most peas on their plate?")—data analysis. They are learning about probability (e.g. "You think it will rain today? Why do you think that?"). Reflect on whether you need to reframe your own understanding of mathematics learning in early childhood.
- Reflect on whether you need to improve your mathematical content knowledge. Find a text that explains the learning trajectory for mathematics in early childhood.
- Remember that learning occurs when the child's thinking, the child's body, the child's feelings, objects in the learning environment, and a problem that is meaningful to the child, all intra-act to support learning.
- Encourage learning by creating a learning environment that sets up opportunities for exploration and conversation.

- Make explicit connections between the words and physical experiences to help children associate mathematical language with *feeling* these concepts concurrently in their bodies.
- Have fun!

References

Anders, Y., & Rossbach, H. G. (2015). 'Preschool Teachers' Sensitivity to Mathematics in Children's Play: The Influence of Math-Related School Experiences, Emotional Attitudes, and Pedagogical Beliefs.' *Journal of Research in Childhood Education, 29*(3), 305–322. https://doi.org/10.1080/02568543.2015.1040564.

Bailey, D., Siegler, R., & Geary, D. (2014). Early predictors of middle school fraction knowledge. *Developmental Science, 17*(5), 775–785.

Barad, K. M. (2007). *Meeting the universe halfway: Quantum physics and the entanglement of matter and meaning*. Durham, NC: Duke University Press.

Baroody, A. J., Clements, D. H., & Sarama, J. (2019). Teaching and learning mathematics in early childhood programs. In C. Brown, M. B. McMullen, & N. File (Eds.), *Handbook of early childhood care and education* (pp. 329–353). Hoboken, NJ: Wiley Blackwell Publishing.

Bennett, J. (2005). Curriculum issues in national policy-making. *European Early Childhood Education Research Journal, 13*(2), 5–23.

Björklund, C. (2013). Matematiklärande för de allra yngsta. [Mathematical learning for the very youngest]. In Holmqvist Olander (Eds.). *Learning study in preschool*. Lund, Sweden: Studentlitteratur.

Broström, S. (2017). A dynamic learning concept in early years' education: A possible way to prevent schoolification. *International Journal of Early Years Education, 25*(1), 3–15. https://doi.org/10.1080/09669760.2016.1270196

Claessenn, A., & Engel, M. (2013). 'How Important Is Where You Start? Early Mathematics Knowledge and Later School Success.' Teachers College Record. The voice of scholarship in education. V 115 (6).

Clements, D., & Sarama, J. (2014). *Learning and teaching early math: The learning trajectories approach*. New York: Routledge.

Ehrlich, S. B., Levine, S. C., & Goldin-Meadow, S. (2006). The importance of gesture in children's spatial reasoning. *Developmental Psychology, 42*(6), 1259–1268.

Elia, I., Gagatsis, A., & van den Heuvel-Panhuizen, M. (2014). The role of gestures in making connections between space and shape aspects and their verbal representations in the early years: findings from a case study. *Mathematics Education Research Journal, 26*(4), 735–761. https://doi.org/10.1007/s13394-013-0104-5

Fisher, K., Hirsh-Pasek, K., & Golinkoff, M. (2012). Fostering mathematical thinking through playful learning. In E. Reese & S. P. Segate (Eds.), *Contemporary Debates on Child Development and Education* (pp. 81–92). New York: Routledge.

Franzén, K. (2014). Under threes mathematical learning: The teacher's perspective. *Early Years. An International Journal, 34*(3), 241–254.

Franzén, K. (2015a). Under threes mathematical learning. *European Early Childhood Education Research Journal.* https://doi.org/10.1080/1350293X.2014.970855

Franzén, K. (2015b). Being a tour guide or travel companion on the children's knowledge journey. *Early Child Development and Care, 185*(11–12), 1928–1943.

Gelman, R., & Gallistel, R. C. (1978). *The child's understanding of number.* Cambridge, MA: Harvard University Press.

Hedge, K., & Cohrssen, C. (2019). Between the red and yellow windows: A fine-grained focus on supporting children's spatial thinking during play. *SAGE Open.* https://doi.org/10.1177/2158244019809551

Horton, J., & Kraftl, P. (2006). What else? Some more ways of thinking and doing 'children's geographies. *Children's Geographies, 4*(1), 69–95.

Lenz Taguchi, H. (2010). Rethinking pedagogical practices in early childhood education: A multidimensional approach to learning and inclusion. In N. Yelland (Ed.), *Contemporary perspectives on early childhood education* (pp. 14–32). Maidenhead, Berkshire: Open University Press.

Lenz Taguchi, H. (2011). Investigation learning, participation and becoming in early childhood practices with a relational materialist approach. *Global Studies of Childhood, 1*(1), 36–50.

Macedonia, M. (2019, October 1). Embodied learning: why at school the mind needs the body. *Frontiers in Psychology, 10,* 2098.

McCray, J., & Chen, J.-Q. (2012). Pedagogical content knowledge for preschool mathematics: Construct validity of a new teacher interview. *Journal of Research in Childhood Education, 26*(3), 291–307.

Merrell, F. (2003). *Sensing corporeally: Toward a posthuman understanding.* Toronto, Canada: University of Toronto Press.

Nichols, J., Levay, K., O'Neil, M., & Volmert, A. (2019). *Reframing early math learning.* Retrieved from http://frameworksinstitute.org/assets/files/Early%20Math/early-math-learning-mm-2019.pdf

Ontario Ministry of Education. (2014). *Paying attention to spatial reasoning K-12: Support document for 'paying attention to mathematics education'.* http://www.edu.gov.on.ca/eng/literacynumeracy/lnspayingattention.pdf

Opperman, E., Anders, Y., & Hachfeld, A. (2016). The influence of preschool teacher's content knowledge and mathematical ability beliefs on their sensitivity to mathematics in children's play. *Teaching and Teacher Education, 58*, 174–184.

Owens, K. (2015). *Introduction: Visuospatial reasoning in context. In Visuospatial reasoning: An Ecocultural Perspective for Space, Geometry and Measurement Education* (pp. 1–16). Springer.

Palmer, A. (2011). *Hur blir man matematisk? Att skapa nya relationer till matematik och genus i arbetet med yngre barn.* [How to become mathematical? creating new relationships with mathematics, and gender in the work with young children]. Stockholm, Sweden: Liber.

Pólya, G. (2004). *How to solve it: A new aspect of mathematical method.* Princeton, NJ: Princeton University Press.

Rautio, P. (2014). Children who carry stones in their pockets: On autotelic material practices in everyday life. *Children's Geographies, 11*(4), 394–408.

SFS. (2010:800). *Skollag (The Education Act).* Stockholm, Sweden: Skolverket.

Sheridan, S., Pramling Samuelsson, I., & Johansson, E. (2009). *Barns tidiga lärande: En tvärsnittsstudie om förskolan som miljö för barns lärande* [Children's early learning: A cross-sectional study of preschool as environment for children's learning]. (Göteborgs Studies in Educational Sciences 284). Göteborg, Sweden: Göteborgs University.

Skolverket [The Swedish National Agency for Education]. (2010). *Läroplan för förskolan, Lpfö 98*, Rev. ed. [Curriculum for preschool Lpfö98]. Stockholm, Sweden: Skolverket. [The Swedish National Agency for Education].

SNAE, Swedish National Agency for Education. 2018. Läroplan för förskolan 2018. Stockholm: Skolverket.

Watts, T., Duncan, G., Siegler, R., & Davies-Kean, P. (2014). 'What's Past Is Prologue: Relations Between Early Mathematics Knowledge and High School Achievement. 'Educational Researcher. https://doi.org/10.3102/0013189X14553660.

Yelland, N. (2010). Extended possibilities and practices in early childhood education. In N. Yelland (Ed.), *Contemporary perspectives on early childhood education.* Maidenhead, UK: Open University Press.

11

The Mechanics of Interaction in Early Childhood STEAM

Amelia Church and Caroline Cohrssen

Introduction

Social interaction is a locus for learning, because children learn about their world through interactions with others. From birth, infants encounter the patterns and principles of turn-taking in conversation, and as their language develops, children take a more active role in initiating and sustaining conversations with parents, siblings, peers and teachers. Participation in conversation enables the exploration of concepts, and extending topics through sequences of questions, answers and attuned responses allows us to see what children know and build on this knowledge. The quality of conversations with children, both at home with

A. Church (✉)
The University of Melbourne, Parkville, VIC, Australia
e-mail: achurch@unimelb.edu.au

C. Cohrssen
The University of Hong Kong, Hong Kong, SAR, China
e-mail: cohrssen@hku.hk

C. Cohrssen, S. Garvis (eds.), *Embedding STEAM in Early Childhood Education and Care*, https://doi.org/10.1007/978-3-030-65624-9_11

parents and with peers and teachers in early childhood settings, has a great deal of influence on the quality of early learning.

Research in early childhood education has shown that high-quality interactions start with joint attention. Joint attention happens when two people pay attention to the same object or actions, orienting to each other's perspective on this same object. Joint attention can be achieved through a range of verbal or non-verbal actions. For example, an infant may turn her head away to indicate 'enough' and the responsive adult recognizes this action as communicating a particular need. A toddler may point at an object and the adult responds by handing the object to the child, commenting on some property of the object or the child's attention. Sometimes, conversations are a mixture of verbal and non-verbal: a child may dance in circles when exploring the concept of 'round' and the adult may comment, 'You are dancing in circles, around and around'. In each example, the adult and the child are both paying attention to the same object or action.

Joint attention is a prerequisite for learning in any environment—at home with parents or in early childhood settings—but we need to do more than simply acknowledge the same object for children's cognitive growth. Extended interactions with children are opportunities for adults to support concept development, to provide feedback that consolidates and extends learning, and to model advanced language (Pianta, La Paro, & Hamre, 2008). This happens when adults ask questions or prompt discussion that builds on what children already know, stimulates curiosity, and supports higher-order thinking and concept transferral. The role of the early childhood teacher is to respond attentively to children and to model the language that is associated with the child's thinking or behaviour. For example, the language of mathematics has been shown to be vital in fostering acquisition of language and meta-cognitive abilities which, in turn, supports the development of mathematical thinking (Warren & de Vries, 2009).

Research has found, however, that early childhood teachers may require support with enacting reciprocal learning interactions with children. Large-scale studies in the United Kingdom (EPPE; Sylva, Melhuish, Sammons, Siraj-Blatchford, & Taggart, 2010), Australia (E4Kids; Tayler, 2016) and the United States (La Paro et al., 2009) reveal that early childhood settings typically score well on providing emotional support for children and on providing opportunities for children to keep busy, but that cognitively stimulating opportunities—created in extended conversations between children and teachers—are infrequent. The reasons for this are complex and numerous. They include the qualifications and experience of staff, teachers' attitudes and beliefs, policy imperatives, time demands, staff ratios and access to professional learning.

Professional learning is the purpose of this book, specifically in relation to how teachers and parents can support Science, Technology, Engineering, Art and Mathematics (STEAM) in early learning contexts. Each chapter in this book provides illustrations of how to create opportunities for learning in play-based STEM or STEAM activities or narratives. This chapter provides foundational knowledge of the elements of interaction itself by highlighting some of the building blocks of what teachers refer to as cognitively challenging conversations (Durden & Dangel, 2008; Massey, 2004) or sustained shared thinking (Siraj & Asani, 2015; Siraj, Kingston, & Melhuish, 2015).

If we want to know how talking with children advances their learning and development, it helps to understand how talk is structured and organized. We intuitively understand that effective communication is key to effective concept development in interactions with young children. For professional learning, however, we want to move past intuition and experience. We want to draw on substantive evidence of how learning interactions improve children's outcomes. When we can identify the 'how' of effective talk-in-interaction, we can include this in our interactions with children in purposeful and responsive conversations. Although many methodologies are concerned with the intricacies of talk, this chapter relies on the transparency of conversation analysis (CA) to tease apart interactions between children and teachers.

The examples in this chapter—extracts of published research in conversation analysis—draw from activities focused on spatial thinking, measurement, particles and matter, and digital literacy, but the insights apply to interactions between teachers and children more broadly. To show how an understanding of the mechanics of interaction can inform intentional teaching practices, we will explain findings from CA research. In this chapter we will (1) introduce the main features of talk-in-interaction (i.e. what?), (2) explain why an understanding of these features is useful for teaching and learning in early childhood education and care (ECEC) (i.e. why?) and (3) provide examples of interaction between teachers and four-year-old children which support concept development (i.e. how?).

Conversation analysts observe naturally occurring interactions and then transcribe and analyse this video-recorded data with 'unmotivated looking'. The analyst is not looking for categories of actions (cf. Land, Tyminski, & Drake, 2019), but rather to see how actions relate to one another in the ongoing sequences of talk. We can see what happens in real-life contexts: we pay attention to the very same things that children, parents and teachers are noticing. This approach does not make assumptions about what might be relevant to children's exploration of STEAM concepts. Instead, conversation analysis provides a transparent tool for looking at talk. This enables us to see where particular practices lead during typical learning-in-interaction that happens every day in early childhood settings. By enabling us to see interactions that support learning, we can try these interactions ourselves with the children in our classrooms or at home.

What Do We Need to Know About the Structure of Interactions?

The rules that govern the organization of talk are important in conversation analysis. They reveal the 'how' of talk-in-interaction. Sacks, Schegloff, and Jefferson (1974) set out the foundational rules of conversation: (1) turns at talk occur one at a time; (2) speakers take turns, that is, one person speaks after the other; and (3) turn-taking is repeated. This appears simple, but it is this nextness, the 'one turn after another' organization of talk as a social practice, that allows us to make sense of one another. When the listener demonstrates they understood the speaker's previous turn(s), intersubjectivity—shared meaning—is achieved. This common focus enables people to manage, negotiate and track shared meaning, as 'next-turn proof procedure' (Hutchby & Woofitt, 1998, p. 15; Schegloff

& Sacks, 1973). In other words, what happens *next* reveals an understanding of prior turns. Sometimes, turns are actions or gestures. For example, when describing the shape of a roof, a child may use their hands to show the shape (Hedge & Cohrssen, 2019). At times, children initiate the talk and the role of the adult is to provide feedback that consolidates what the child knows and to extend that knowledge a little further (Church & Bateman, 2019). With each child's turn, we can see what they understood from the prior talk and how they mobilize this understanding in their response to it in the next action (Stivers & Rossano, 2010). It is the nextness of these turns at talk that provide adults with evidence of children's understanding and orientation to the concept. This evidence provides opportunities for in-the-moment assessment.

If we think of turns at talk as achieving actions (see Enfield & Sidnell, 2017), we can see that in a sequence there are relevant *next* actions. A question should usually be followed by an answer. There are all sorts of things that speakers do to orient to this rule, including repeating the question if no answer is forthcoming. Indeed, the third turn in a sequence of talk is where pedagogy resides (Cohrssen & Church, 2017; Lee, 2007; Park, 2015) by providing feedback, acknowledging the learner's contribution, and reorienting the group to the relevant *next* action. However, teaching and learning are not limited to three-part sequences. They usually unfold in a *series* of turns, where the sequence has multiple turns and builds cumulatively.

In the following example from Bateman's (2013) study of intersubjectivity in early learning environments in New Zealand, the teacher is exploring concepts of measurement and impermanence with two three-year-old boys. The teacher (ECT) is sitting on the side of the sandpit, where the boys (Fred and Levi) are pouring buckets of water into the sand. Free play in the sandpit provides rich opportunities for STEAM learning. This transcript excerpt uses CA transcription convention (Jefferson, 2004), and the punctuation is used to try to capture how the talk is done (emphasis, overlap, pauses and so on; see Appendix).

Extract 1

```
79 ECT:  the:re we go:=look how its flowing down=but what's
         happened? (0.6)
80       where's the water gone.
81 Fred: gone. ((holds palms of hands out to sides))
82       (2.9)
83 Levi: its not working. ((runs to get more water))
84 ECT:  its not working anymore. why do you think it's not
         working anymore.
85       ((continues digging sand))
86 Fred: coz it's not.((digs with teacher))
87 ECT:  coz it's no::t.
88 Fred: [no]
89 ECT:  [but] why isn't it working anymore. do you know why?
90       (6.5)
91 Fra:  ((runs to get more water))
```

This episode continues for some time with the boys continuing to pour water into the sand and the teacher persisting with prompts for hypotheses ('why do you think…'). The boys engage with these prompts, proposing the water has gone 'under' and 'through that hole'. The teacher's repetition serves to acknowledge each observation made by Fred and encourage ongoing exploration of the process. Bateman (2013, p. 280) notes that 'by indicating that there is a problem worth investigating in the sandpit, [the teacher] is opening the topic up for further discussion with the children.' We see that building sequences of interaction facilitates the exploration of ideas and concepts and that the teacher's turns shape children's contributions to the ongoing activity (Walsh, 2011). It is the extended sequence of question and answers that creates the opportunity for scientific inquiry.

In Extract 1, we saw that the boys' collecting, carrying and pouring of water was embedded in their experience of measurement, fluidity and infiltration. That is, the sequence of talk involved not only utterances from the teacher and children but also accompanying gesture (e.g. open palm of hands for 'gone') and interaction with resources (e.g. carrying the bucket of water). Talk-in-interaction is a term that aims to capture the

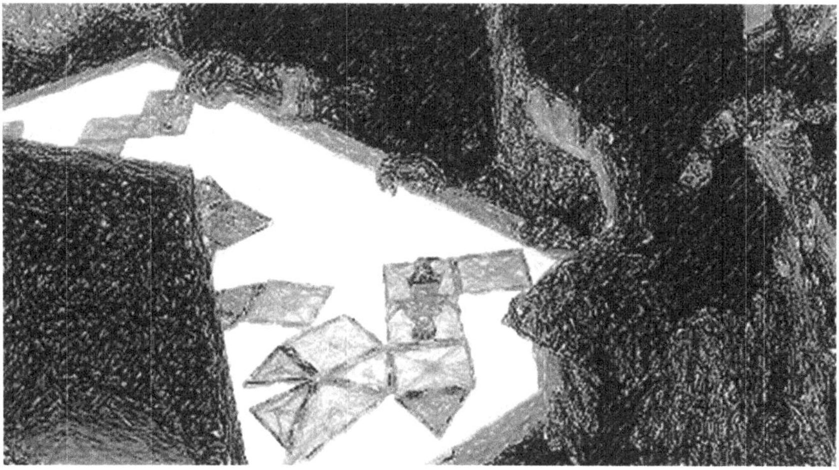

Fig. 11.1 'What's the opposite of this block?'

dynamic and multimodal nature of conversations as talk encompasses all elements of face-to-face communication: words, intonation and emphasis, pauses, eye gaze, gesture and embodied action. In fact, the language we use to express mathematical and scientific concepts is often embodied (e.g. 'larger', 'closer', measured with arm width or gesture, and other deictic pronouns such as 'this', 'that' with pointing or head nods). For example, in Extract 2, the teacher is drawing children's attention to spatial orientation of shapes on a light table (see Fig. 11.1), touching the shapes simultaneously as he speaks (Hedge & Cohrssen, 2019, p. 6):

Extract 2

```
89 TEA:    look. (0.4) Keira is on this one (0.6) what's the-
           what's the
90         opposite of that one.
           on 'this' and 'that' teacher touches block closest
           to centre
91         (0.8)
92 TEA:    what's the opposite of this block.
```

Embodiment is also relevant to the exploration and expression of ideas of younger children as they orient themselves to the world around them. Infants and toddlers may turn their bodies away from food to indicate 'enough' and use a full range of gestures and eye gaze to engage in early learning. Very young children use pointing (Filipi, 2009) and holding objects into the sightline of adults targeted as an audience to label, request, seek information and claim objects for their experiential play (Kidwell, 2011). Older children also use the body to explore scientific and spatial concepts. For example, Hoey, DeLiema, Chen, and Flood (2018) show how children aged five to six years use their bodies to enact the states of matter by 'being' particles during a science lesson (see Hoey et al., 2018, p. 14, Fig. 2). In the activity, children are required to map the position of their own body in the room in relation to each other to see how particles (i.e. their relative positions) are mapped on the screen projection in front of the class.

Why Pay Close Attention to How Talk Is Done?

Close attention to the mechanics of interaction reveals the systematic particularities and predictability of turns at talk. The timing, interdependency and adaptability of the mechanics of talk are complex yet often seamless. For example, overlapping speech is rarely done, as speakers reliably time their next turn within 0.2 of a second of the other speaker finishing (see Enfield, 2017 for review). Children are enculturated into this sensitivity to the speech of others from interactions with parents and other caregivers from birth.

As a methodological approach, conversation analysis reveals the details of talk-in-interaction which we would not otherwise remember (see Sacks, 1984; Stokoe, 2011). Recalling our conversations with children does not capture how topics are realized and talked into being, and we are not always able to imagine what children might do or say. Studying classroom interactions (i.e. through recording, transcribing and analysing the real thing) reveals a perspective that—however experienced we are in working with young children—is often not obvious. Using data from CA—or other research that presents the original data for reflection—allows us to specify how learning opportunities are created.

For example, in Extract 3 (Church & Bateman, 2019, p. 271), Max—aged four years—proposes that the concept and application of time can, in fact, be negotiated:

Extract 3

```
01 ECT: "no one (.) can stop time you know."
02      (2.2)
03 Max: people can stop the time. (0.3) they can just- (0.5)
04      turn the clock and hold it where it is (.) and then
05      they- (0.2) (yeah) (.) people allow them to have time?
06      (0.5)
07 ECT: (well). maybe we could !try that one day. (1.4) and
08      see what happens.
```

Capturing the detail of these learning sequences allows us to see that interaction is a *collaborative* endeavour. We are not prioritizing what the teacher does, or isolating children's responses, but rather concerned with how children and teachers build on each contribution to the ongoing talk as a *jointly coordinated activity*. Teaching and learning are viewed as a social process, and all participants have a role in both the process (all the elements of the talk) and the product (what the interactions lead to in terms of learning). If we return to the example from Church and Bateman (2019, p. 271), we see that the exploration of the concept of time is only made possible through the teacher AND children co-constructing the extended sequence.

Extract 4

```
07 ECT: (well). maybe we could !try that one day. (1.4) and
08      see what happens.
09      (0.6)
10 TCH: what will happen to the rest of (0.6) all the other
11      children if we put the clock back.
12      (0.8)
13 Max: um (.) be too much time (.) no one could
14      [be-]    [ °xxx ° ]
15 Ali: [and](.) [and we'll] be muddled up and we'll have
16      dinner at- (1.6)<!breakfast ti:me.>=
```

The discussion continues with all four children sitting with the teacher as she reads a book about a child going in search of time. The sequence is built, however, with the teacher designing questions to elicit theories from the children and the children not only collaborating in their responses to these prompts, but also recycling and extending suggestions made by other children in the group (see Walsh & Li, 2013, for further discussion of successful elicitation in classroom interaction). The details of *how* the talk is done prove useful for reflection and development—for teachers and parents alike. This means that we can base our understandings of effective interactions on evidence of what actually happens and how sustained-shared thinking is actually built.

How Can Teachers Extend Interactions in STEAM?

We have made a case for paying close attention to naturalistic data for professional learning. Now let's look at the detail of interactions between children and teachers that focus on STEAM learning. Each chapter in this book details different aspects or concepts explored in early childhood education and so here we will illustrate the 'how' of three features of high-quality learning interactions: (1) elements of question design; (2) how pauses can be used to enable opportunities for thinking; and (3) how responsiveness can be (collaboratively) achieved.

How we design questions determines the range of possible next actions for children. The design of questions is an important part of intentional teaching. Teachers aim to use 'open-ended questions' (e.g. 'why do you think…?') rather than closed questions (e.g. 'what colour is that?') to encourage children's exploration and agency (see Siraj-Blatchford & Manni, 2008). Using an 'open-ended' question, however, does not necessarily lead to learning-in-interaction. Researchers have shown that particular question prefaces actually increase the likelihood of children offering their ideas or opinions. Epistemic asymmetry (teachers having more knowledge than children) is inevitable in classrooms (see Gardner, 2019), but lessening this asymmetry creates space for exploration of ideas (see Walsh & Li, 2013, on conversational space for learning). Sandra Houen, for example, found that when teachers used 'I wonder' formulations, this less-knowing stance adopted by the teacher was 'more successful than the explicit "wh" question design in gaining a response' (Houen, Danby, Farrell, & Thorpe, 2016, p. 75).

In Extract 5, the teacher (ECT) and a small group of four-year-old children have been looking at an image of a lady beetle on a computer (Houen et al., 2016, p. 74). They have talked about and counted the six legs. Here, the teacher draws the children's attention to the lady beetle's antennae:

Extract 5

```
156 ECT:  =what's tha:t. ((gazes at Mena))
156       (1.6)
157 ECT:  >I wonder what< that is.
158       (0.6)
159 Mena: ey::es. ((gazes at teacher))
160 ECT:  ey:es!¿=
161       =do you think it might be eye:s.
162       (2.2)((all gaze at screen))
163 Rory: an:tennaes ((looking at teacher))"
164       (0.4)
165 ECT:  >you think it might be< antennaes Rory.=
166 Mena: =this is-=
167 ECT:  =>what do they use< antennaes for.((gazes at Rory))
```

In this extract, we see that the teacher was seeking a response from Mena in particular by looking at her while asking the question 'What's that?'. The teacher gives Mena time to consider and provide a response (note the 1.6-second pause at line 158) before reformulating the question as an invitation to volunteer ideas rather than provide the correct (known by the teacher) answer. We can see that this is effective, because in response to the 'I wonder' formulation, Mena offers the suggestion 'eyes'.

The subsequent neutral turn from the teacher (i.e. the teacher does not evaluate 'eyes') invites further suppositions from the group, and Rory (line 163) provides the correct anatomical noun for the antennae. Houen et al. (2016) propose that it is the formulation 'I wonder' that encourages Mena to offer her ideas in the group discussion. This claim can be made based on the evidence within the interaction itself: Mena does not respond to the initial request for factual information (Margutti, 2007), but she does attempt an answer following the teacher's re-framing of the question as 'I wonder what that is', and in addition, she is sufficiently encouraged by the re-design of the question to continue to offer her ideas with the group (see the halted turn in line 166).

Extract 5 illustrates the importance of question design in eliciting evidence of what children already know. It also provides an example of how pauses can be used in STEAM pedagogy to facilitate learning, by allowing children time to consider the problem posed (i.e. Rowe's definition of 'wait time'; 1986, 1987) and formulate a relevant response. Cohrssen, Church, and Tayler (2014a, 2014b) have shown the importance of teachers' pausing at decisive points in the interaction when engaged in mathematics activities. Creating space in the interaction (Walsh & Li, 2013), in addition to providing optimal opportunities for children to respond, also allows teachers to assess children's mathematical knowledge. As a result, they are equipped to design subsequent turns at talk that align with children's current knowledge and extend concepts through targeted use of questions and reinforcing content (i.e. scaffolding learning).

In Extract 6 (Cohrssen et al., 2014b, p. 97), the teacher (ECT) is playing a game with three children, aged between four and five years, all of whom have English as a second language, where each child, in turn, rolls a die, counts the number of dots, then chooses the corresponding number of objects from the counters in the centre of the table. Just prior to this excerpt, Halla has rolled the die, which has landed with six on top.

Extract 6

46 Hal: ((*looks at teacher*))
47 ECT: ((*nods*))
48 Hal: ((*pushes the die in direction of teacher, then looks at teacher*))
49 ECT: °how much di=did you roll?°((*looks at Halla, then passes the die back*
50 *to her*))
51 Hal: ((*touches the surface of the die repeatedly with an index finger,*
52 *then looks up at teacher who is watching*))
53 (2.0)
54 ECT: shall we touch them? (.) together? ((*touches the die*))
55 (0:2)
56 Hal: ((*nods, puts her head to one side*))

In this sequence, the teacher provides a range of encouragements for Halla to identify the number on the die, including the embodied go-ahead (the nod in line 47) in response to Halla's unspoken request for assistance (looking at the teacher in line 46). Notably, in line 53, the teacher waits—two seconds is an audibly extended pause—for Halla to count the dots and provide an answer. When Halla returns her gaze to the teacher, the teacher then provides an alternative next action, inviting the group to count 'together' (line 54), which is immediately agreed to by Halla. Following Halla's affirmation, the teacher invites the other two children to join in the counting, 'let's help her count them' (line 57).

Extract 7

57 ECT: let's help her. count them. ready? you touch them.
58 Hal: [((*touches each dot in time with each number spoken/chant by*
59 *Lukaz and Johanna*))]
60 Luk: [one, two, three, four, five, six]
61 Joa: [one, two, three, four, five, six]
62 ECT: [((*nods in time with the count*))]

```
63 Hal: ((looks at teacher))
64 ECT: six? ((nodding))
65 ECT: c'n you count out six pieces? ((points to box))
66 Luk: [(speech inaudible due to classroom noise)]
67 Joa: [(speech inaudible due to classroom noise)]
68 Hal: [((draws six counters towards her, one at a
        time))] (9:4)
69 ECT: okay, ((picks up the die))
```

It becomes clear once Halla selects six objects that she understands one-to-one correspondence. This was not evident at the beginning of the sequence, but the teacher does not treat the absence of relevant next action (line 48) as an indication that Halla does not know the answer. English is Halla's second language; she is also quite shy and soft-spoken. The extended pauses and multiple opportunities created by the teacher enable Halla to display her knowledge, engage in the activity as part of a group, and build confidence and self-efficacy in the game. Assessment for learning at this moment requires extended pauses—including allowing time for Hall a to select six counters from the centre of the table—to create opportunities for Halla to demonstrate her mathematical knowledge.

The pauses also provide emotional support for the children involved in the activity. The longer sequence from which the extract above was taken shows the teacher moderating her requests for information based on the individual needs—that is cognitive, linguistic and social—of each child (see Cohrssen et al., 2014b). Creating spaces for a range of possible answer actions from children through the purposeful use of extended pauses is intentional teaching which respects and recognizes the abilities of each child (Church & Bateman, 2019; Walsh & Li, 2013). Responsive interactions build foundations of responsive and respectful relationships (Australian Early Years Learning Framework, DEEWR, 2009) with peers and adults alike. This responsivity focuses our attention on the collaborative nature of talk; not so much what the teachers say, but what the children and teacher do together.

The next example in Extract 8 provides a delightful illustration of how intentional teaching can be built into play-based learning. Here is an activity of high-jinks: three boys are playing around the school and grounds they have built with blocks, part of a six-week inquiry project (see Cohrssen, de Quadros-Wander, Page, & Klarin, 2017 for details). Learning experiences in this project were built around 'starting school' and children sketched, photographed, mapped and designed 2D and 3D representations of the school they would attend the following year. This excerpt is taken from an activity where the teacher's aim was to consolidate the children's use of spatial language (e.g. in, on, under, up, down, beside, between, in front of, behind, near, far, opposite, left and right).

The boys are laughing, moving around the blocks (the school) on the table, making jokes and being silly. The teacher—and here is the responsive design of his talk—joins the children in their activity. He is pretending to be a delivery man on the phone at the school gates, asking children to put pictures of objects in specific places in the 3D/block plan of the school. This has been purposefully planned to enable the teacher to assess the boys' mastery of concepts of relative position (i.e. on top of, underneath etc.). The responsivity of the talk is found in the teacher's decision to join in the silliness of the boys' already established play (evident in the laughter, use of exaggerated gestures and making jokes), by giving nonsensical directions for these objects. This extract from Hedge and Cohrssen (2019, p. 7; talk from two other boys off camera not included here) begins with a question designed by the teacher to elicit assessment of Humphrey's understanding of 'in between' that functions as a request from the delivery man providing instructions of where objects are to be distributed.

Extract 8

```
05 ECT: [I havva delivery=c'n you tell wo=this=is,
06 HUM: computer!
...
09 TEA: computer, well done, now this computer (.) [needs ta
        go::,] on top
10      of the::: red no=in between a red an yellow window, please.
11      ((passes card to Humphrey))
12 HUM: (okay,) ((walks around the table to a position opposite
        the teacher))
..
```

19 TEA: [in=between, a red an yellow window.
20 (4.0)
21 HUM: ((*walks around the table, looking at the sides of the block*
22 *construction of a school building*))
23 GAR: it's=supposed tah go hee-yah! ((*gestures towards building*))
24 HUM: red=an=yellow window.
25 (3.0)
26 HUM: ((*walks back around the table then places card on top of red and*
27 *yellow window*))

This short sequence tells us many things about four-and-a-half-year-old Humphrey's spatial reasoning, given that he places the picture of the computer *on top of* rather than *in between* the windows. We provide the extract to illustrate how responsive and respectful practice can be achieved in that the teacher has maintained the learning goals (reinforcing learning of spatial orientation) but has managed to build this pedagogy into the boys' already established play (the teacher goes on to join in the jokes, asking 'Humphrey Highpants' to deliver objects etc.). Paying close attention to the sequences of actions allows us to see how STEAM interactions can be done.

Conclusion

Social interaction is the locus of learning in early childhood, and opportunities for STEAM learning occur throughout the day, across activities, indoors, outdoors, with peers, with teachers and with family members. In this chapter, we have illustrated the fine-grained detail of high-quality interactions that support learning. We have shone a spotlight on the what, how and why of effective adult-child interactions. We have used examples of talk-in-interaction during science and mathematics activities in order to show how teachers intentionally build on immediately prior actions to support children's concept acquisition and language learning. We hope that by making specific interactional characteristics of effective talk-in-interaction visible, early childhood teachers and parents will feel encouraged to employ these strategies in their interactions with children for purposeful, evidence-based and responsive opportunities for learning.

Appendix: Transcription Conventions

The transcription conventions used in this article follow the original work of Sacks et al. (1974).

.	falling intonation
,	slightly rising or continuing intonation
?	rising intonation
¿	intonation that rises more than a comma but less than a question mark
::	lengthened syllable
↓	sharp fall in pitch
↑	sharp rise in pitch
Bold	emphasis
CAP	increased volume
[ ]	overlapping talk
( )	unintelligible stretch
(0.5)	length of silence in tenths of a second
<	increase in tempo, rushed stretch of talk
< >	slower tempo
hh	audible outbreath
.hh	audible inbreath
[°]	talk that is quieter than the surrounding talk
$	spoken while smiling
(( ))	description of accompanying behaviour

References

Church, A., & Bateman, A. (2019). Children's right to participate: How can teachers extend child-initiated learning sequences. *International Journal of Early Childhood, 51*(3), 265–281.

Cohrssen, C., & Church, A. (2017). The 'how' of high quality child-educator interactions in play-based mathematics. In A. Bateman & A. Church (Eds.), *Children's knowledge-in-interaction: studies in conversation analysis*. Singapore: Springer.

Cohrssen, C., Church, A., & Tayler, C. (2014a). Purposeful pauses: Teacher talk during early childhood mathematics activities. *International Journal of Early Years Education, 22*(2), 169–183.

Cohrssen, C., Church, A., & Tayler, C. (2014b). Pausing for learning. *Australasian Journal of Early Childhood, 39*(4), 95–102.

Cohrssen, C., de Quadros-Wander, B., Page, J., & Klarin, S. (2017). Between the big trees: A project-based approach to investigating shape and spatial thinking in a kindergarten program. *Australasian Journal of Early Childhood, 42*(1), 94–104.

Department of Education, Employment and Workplace Relations (DEEWR) for the Council of Australian Governments. (2009). *Belonging, being and becoming: The early years learning framework for Australia*. Canberra, ACT: Commonwealth of Australia.

Durden, T., & Dangel, D. R. (2008). Teacher-involved conversations with young children during small group activity. *Early Years, 28*(3), 251–266.

Enfield, N. J. (2017). *How we talk: The inner workings of conversation*. New York: Basic Books.

Enfield, N. J., & Sidnell, J. (2017). *The concept of action*. Cambridge, UK: Cambridge University Press.

Filipi, A. (2009). *Toddler and parent interaction: the organization of gaze, pointing and vocalization*. Amsterdam: John Benjamins.

Gardner, R. (2019). Classroom interaction research: The state of the art. *Research on Language and Social Interaction, 52*(3), 212–226.

Hedge, K., & Cohrssen, C. (2019). Between the red and yellow windows: A fine-grained focus on supporting children's spatial thinking during play. *SAGE Open, 9*(1), 1–11. https://doi.org/10.1177/2158244019829551

Hoey, E. M., DeLiema, D., Chen, R. S. Y., & Flood, V. (2018). Imitation in children's locomotor play. *Research on Children and Social Interaction, 2*(1), 1–24.

Houen, S., Danby, S., Farrell, A., & Thorpe, K. (2016). 'I wonder what you know…': Teachers designing requests for factual information. *Teaching and Teacher Education, 59*, 68–78.

Hutchby, I., & Wooffitt, R. (1998). *Conversation analysis: Principles, practices and applications*. Cambridge, UK: Polity Press.

Jefferson, G. (2004). Glossary of transcript symbols with an introduction. In G.H. Lerner (Ed.) *Conversation Analysis: Studies from the First Generation*, (pp.13–31). Amsterdam / Philadelphia, John Benjamins.

Kidwell, M. (2011). Epistemics and embodiment in the interactions of very young children. In T. Stivers, L. Mondada, & J. Steensig (Eds.), *The morality of knowledge in conversation* (pp. 257–284). Cambridge, UK: Cambridge University Press.

La Paro, B., Hamre, K., Locasale-Crouch, J., Pianta, R., Bryant, D., Early, D., et al. (2009). Quality in kindergarten classrooms: Observational evidence for the need to increase children's learning opportunities in early education classrooms. *Early Education and Development, 20*(4), 657–692.

Land, T. J., Tyminski, A. M., & Drake, C. (2019). Examining aspects of teachers' posing of problems in response to children's mathematical thinking. *Journal of Mathematics Teacher Education, 22*(4), 331–353.

Lee, Y.-A. (2007). Third turn position in teacher talk: Contingency and the work of teaching. *Journal of Pragmatics, 39*, 180–206.

Margutti, P. (2007). "Are you human beings?" Order and knowledge construction through questioning in primary classroom interaction. *Linguistics and Education, 17*(4), 313–346.

Massey, S. (2004). Teacher–child conversation in the preschool classroom. *Early Childhood Education Journal, 31*(4), 227–231.

Park, I. (2015). Or-PREFACED third turn self-repairs in student questions. *Linguistics and Education, 31*(1), 101–114.

Pianta, R. C., La Paro, K. M., & Hamre, B. K. (2008). *Classroom Assessment Scoring System™: Manual K-3*. Paul H Brookes Publishing.

Rowe, M. (1986). Wait time: Slowing down may be a way of speeding up! *Journal of Teacher Education, 37*, 43–50.

Rowe, M. (1987). Using wait time to stimulate inquiry. In W. Wilen (Ed.), *Questions, questioning techniques, and effective teaching* (pp. 95–106). Washington, DC: National Education Association.

Sacks, H. (1984). Notes on methodology. In J. M. Atkinson & J. Heritage (Eds.), *Structures of social action* (pp. 21–27). Cambridge, UK: Cambridge University Press.

Sacks, H., Schegloff, E. A., & Jefferson, G. (1974). A simplest systematics for the organization of turn-taking for conversation. *Language, 50*(4), 696–735.

Schegloff, E. A., & Sacks, H. (1973). Opening up closings. *Semiotica, 8*, 289–327.

Siraj, I., & Asani, R. (2015). The role of sustained shared thinking, play and metacognition in young children's learning'. In S. Robson & S. Quinn (Eds.), *The Routledge international handbook of young children's thinking and understanding*. London: Routledge.

Siraj, I., Kingston, D., & Melhuish, E. (2015). *Assessing quality: Sustained shared thinking and emotional well-being (SSTEW) rating scale*. London: Trentham Books & UCL Press.

Siraj-Blatchford, I., & Manni, L. (2008). "Would you like to tidy up now?" An analysis of adult questioning in the English foundation stage. *Early Years, 28*(1), 5–22.

Stivers, T., & Rossano, F. (2010). Mobilizing response. *Research on Language and Social Interaction, 43*(1), 3–31.

Stokoe, E. (2011) Simulated interaction and communication skills training: The 'Conversation Analytic Roleplay Method'. In Antaki, C.E. (Ed.) *Applied conversation analysis: Changing institutional practices* (pp. 119–139). Basingstoke: Palgrave MacMillan.

Sylva, K., Melhuish, E., Sammons, P., Siraj-Blatchford, I., & Taggart, B. (2010). *Early childhood matters: Evidence from the effective pre-school and primary education project*. London: Routledge.

Tayler, C. (2016). Reforming Australian early childhood education and care provision (2009–2015). *Australasian Journal of Early Childhood, 41*(2), 27–31.

Walsh, S. (2011). *Exploring classroom discourse: language in action*. London: Routledge.

Walsh, S., & Li, L. (2013). Conversations as space for learning. *International Journal of Applied Linguistics, 23*(2), 247–266.

Warren, E. and de Vries, E. (2009) Young Australian Indigenous students' engagement with numeracy: Actions that assist to bridge the gap, *Australian Journal of Education, 53*(2), Article 4.

12

STEM Learning Ecologies: Productive Partnerships Supporting Transitions from Preschool to School Growing a Generation of New Learners

Nicola Yelland

Introduction

Science, Technology, Engineering and Mathematics (STEM) education has become an imperative in education in recent times with the growing realisation that schooling systems require additional strategies and measures to prepare students for life in society which are very different from those in previous eras (Aldemir & Kermani, 2016; Carter, 2016; Early Childhood STEM working group, 2017; Katz, 2010). In Australia this imperative was instigated by the National Science Innovation Agenda (NISA, 2015a), which was the catalyst for the National STEM school education strategy (2016–2026) (NISA, 2015b). The strategy promoted the imperatives for developing mathematical, scientific and digital

N. Yelland (✉)
The University of Melbourne, Parkville, VIC, Australia
e-mail: nyelland@unimelb.edu.au

C. Cohrssen, S. Garvis (eds.), *Embedding STEAM in Early Childhood Education and Care*, https://doi.org/10.1007/978-3-030-65624-9_12

literacy for all Australian children, as well as promoting problem solving, critical analysis and creative thinking skills in authentic investigative learning environments.

The challenge was also taken up by the twenty-first century skills movement (e.g. Partnerships for the 21st century, 2008; Trilling & Fadel, 2009) which emphasised the importance of learning dispositions, and the acquisition of the key learning skills of critical thinking, creativity, collaborations and communication. This interdisciplinary approach could easily have incorporated Arts and become Science, Technology, Engineering, Arts and Mathematics (STEAM), yet the STEM acronym has already become pervasive in its use at this time, and it would seem it is already accepted by many. In fact, the acronym started out as SMET, but it was soon realised that using STEM was more attractive in application (Blackley & Howell, 2015).

In the context of these developments, it was evident that knowledge creation is fundamental for learning, and that it requires building on existing knowledge and applying skills of inquiry to generate new ideas that are relevant in a variety of contexts. For governments and indeed all educators, the problems associated with contemporary learning have been viewed as a science-based issue, and they have been particularly concerned about the lack of interest students have shown to study the traditional science subjects in schools and universities. This concern is clearly manifested in the STEM movement. In the early years (birth to eight years of age) a focus on STEM thinking has been regarded as being a way to advance changes to existing curriculum by, for example, engaging pre-school children in experimentation and inquiry in play-based learning environments (Cohrssen & Page, 2016; Hofer, Farran & Cummings, 2013).

The origin of the acronym (STEM) dates back to 1985, when the Carnegie Foundation had a stated goal to ensure that all children should have the opportunity to participate in and adapt to a society characterised by changing economies. This was linked to the notion that high performing economies had schools that enabled their students to experience authentic experiences via collaborative learning, critical and creative thinking and the sharing of ideas. Since 1985, educational forums have pondered the ways that the sciences can facilitate knowledge creation and technological innovation in order for economies to flourish in the twenty-first century. While there was a recognition of the importance of the traditional disciplines, it was evident that new ways of thinking about them needed to be promoted. Finding the synergies between the science disciplines required reconceptualising interdisciplinary thinking around their core principles and a realisation that this learning had to occur in contexts that were meaningful to learners.

In 2006 a definition of STEM education was generated by a group in the United States:[1]

> STEM Education is an *interdisciplinary* approach to learning where rigorous academic concepts are coupled with *real world* lessons where students apply science, technology, engineering, and mathematics in contexts that make connections between school, community, work, and global enterprise enabling the development of STEM literacy and with it the ability to compete in the new economy. (p. 3) [italics added for emphasis]

There are *two* key aspects to the definition. First, that STEM has an *interdisciplinary* focus and secondly, that experiences should be *authentic* and involve 'real world' applications of knowledge and skills (Blackley & Howell, 2015; Kelley & Knowles, 2016; Yelland, 2020). In schools, STEM education required that teachers integrate the concepts and processes that were usually taught as separate subjects and emphasise the application of knowledge and skills in real-life situations. STEM experiences were not *lessons* in which content was studied, but rather inquiries that originated in wonderings about the world, and posing questions or hypotheses about authentic events.

STEM in Early Childhood

The contribution of early childhood education and home and family environments and their impact on subsequent learning as children progress through schooling systems has long been recognised (e.g. Schweinhart,

[1] Pennsylvania STEM Network, Southwest Region, "Long Range Plan (2009–2018), Plan Summary," (2009) http://www.cmu.edu/gelfand/documents/stem-survey-report-cmu-iu1.pdf

2007). If we are able to build and support twenty-first century STEM education in the early years, we have the potential to grow a generation of new learners who will participate in knowledge creation, skill building and be confident to solve everyday problems in communities of learners.

Play-based pedagogies complement the interdisciplinary STEM approach to teaching and learning (Bennett, Wood & Rogers, 1997; Yelland, 2011). The essential skills and processes that are inherent to science, mathematics, engineering and design, encourage 'learning by doing' in authentic contexts and provide opportunities for intentional teaching when educators are able to effectively scaffold learning and lay the foundations for future possibilities. What is different about learning today is that it is multimodal with the incorporation of digital devices (Yelland, 2018). Learning occurs in linguistic, visual, kinaesthetic, aural and spatial modalities and occurs in both digital and physical spaces—in the home environment, in early learning centres and in school. Taking this into consideration, the conceptualisation adopted here is that

> *STEM education in the early years provides a context for designing active learning ecologies that connect with children's natural curiosity about their world. It engages children in authentic investigations, using critical and creative thinking in systematic ways to build knowledge, acquire skills and cultivate confident dispositions for learning.*

It is widely believed that curriculum and pedagogies in the early years (birth to eight years of age) occur best in play-based contexts, in which teachers prompt and scaffold children's learning (Brenneman, Lange & Nayfield, 2019; French, 2004; Fusaro & Smith, 2018; John, Sibuma, Wunnava, Anggoro & Dubosarsky, 2018). Effective learning scenarios are flexible, and they originate when the teacher observes and responds to individual children's learning needs. However, when compulsory schooling begins there is pressure on teachers to adopt more formal and didactic pedagogies in which more academic skills and knowledge are foregrounded. In Australia, the Early Years Learning Framework (DEEWR, 2009) guides early childhood educators in the preschool years while the National Curriculum guides school-based teachers. In the case study which follows, it is demonstrated that play-based pedagogies are effective in the early years of school, creating STEM learning ecologies that went beyond the walls of Reception classrooms. Reception teachers maintained their professional responsibility to the mandated National Curriculum whilst enacting innovative pedagogical and curriculum designs. They incorporated detailed pedagogical documentation not only to justify their decision-making but also to act as props for re-launching investigations at a later date as the children became more curious about their findings and reflected on them after time had passed.

Case Study: The STEM Bridge Project

The aims of STEM Bridge Project were to create a learning vision for the young children as one in which they would be confident learners who were:

- **Articulate:** Able to communicate their ideas, feelings and discoveries in different ways showing how they understand their life worlds;
- **Respectful:** Work collaboratively with others in communities of practice, listening and questioning and striving for common and multiple goals; and
- **Knowledgeable:** Able to support their plans with accurate and relevant data, and facts that illustrate logical and creative thinking in diverse contexts.

There were 34 participants in this year-long project. They included 14 preschool educators, eight teachers of children aged six years (enrolled in what is known as the Reception year in the Australian state in which the project took place), six principals and six early childhood leaders (ECL) for preschool in each region. Each school principal and ECL participated in the two professional learning workshops held during the year and also attended the on-site pedagogical conversation meetings (one on each site with the researchers) about the project. In the initial one-day introduction workshop session it was evident from conversations that the preschool educators who participated in the project were confident with play-based pedagogical repertoires, but the Reception teachers were more cautious about giving the children the opportunity to lead their learning—given the mandated imperatives of the Australian National Curriculum. At the initial workshop and in the pedagogical conversations that took place on return to the preschools and schools, the educators and teachers at each site collectively decided on a focus for their collaboration and shared this with the research team in their pedagogical documentation. At each of the six sites, the teachers located in the schools created plans and opportunities in which they described wanting to increase opportunities for child-initiated provocations and hands-on explorations that would encourage the children to imagine, explore and use flexible thinking to solve authentic inquiries. One of the major imperatives for the Reception teachers was that they needed these activities to be within the scope of the topics and content mandated in the National Curriculum.

At the initial whole-day professional learning workshop to introduce the project, the general focus for the project design was negotiated between groups of educators and teachers at each of the sites. In this first session, they expressed bold learning desires and recognised they would have to rethink their approaches and pedagogical strategies in order to incorporate more of the children's voice into their learning encounters. In doing so, they wanted a notable shift in creating learning ecologies which went beyond the traditional learning environments in order to stimulate engagement between people, rather than just between teacher and students.

The research questions were:

1. In what ways can preschool educators and Reception teachers **collaborate** to facilitate transitions?
2. Does STEM learning provide a useful **context** to encourage collaborative learning designs for preschool and Reception learners?
3. What **pedagogies** are enabled in this collaboration?

At the same time, the conversations led the educators and teachers to state their educational aspirations for the children.

Focus on Practice

At the centre of this research project, the teachers engaged in critical action research (CAR) (Kemmis & McTaggert, 2007). CAR was selected as the research methodology since it involves being self-reflective in the study of one's practice, which was at the core of this project. It includes an awareness of the ways in which language is used, a recognition of the impact of context and how power relationships impact on practice with the general goal of wanting to improve practice. It emerges from a desire to change, more specifically to improve, practices (pedagogies and curriculum reconceptualisation). Classroom action research deploys qualitative interpretive methods of inquiry and data collection by preschool educators and teachers, here in collaboration with departmental personnel and a university academic as 'participant observers', with decisions being made about how practices might be improved.

The data collection involved all the preschool educators and teachers collating a reflective journal, collecting samples of children's work, and reflecting on and documenting the pedagogical conversations with the university-based academic and department personnel as well as between themselves. This triangulation of data enabled analyses that were shared by all research participants.

The theoretical sampling of the data using the Grounded Theory Approach (Strauss & Corbin, 1990) enabled preschool educators and teachers, department personnel and the researcher to review our ideas and theories as they emerged during the year. In using grounded theory to analyse the data we were able, as researchers, to generate explanations as theory via an inductive analytic process. This meant that the analysis and consideration of themes, via coding, was characterised by a bottom-up approach incorporating the identification of important pedagogical incidents, analysing them and reporting them in the form of narratives and pedagogical documentation to illustrate the children's meaning-making and understandings about the individual topics in each location. The explanations about learning generated by the educators and teachers from their pedagogical documentation were interpretative since they go beyond the semantic meaning associated with pedagogical decision making into the realms of critically considering the ideas, assumptions and conceptualisations which form the basis of the designs for learning process. As a result, the pedagogical documentation collected and produced to reflect on the learning processes was an integral part of data collection and analysis. Further, the narratives created as case studies reflect practical and insightful documentation of the design and learning process of the children for other educators to consider and use as provocations in their own learning ecologies.

Documentation

The role of documentation by the preschool educators and teachers and by the children became important in this project. For the preschool educators and teachers, the data generated became a record of their actions—a diary of events with examples from practice. For the children, participation in the documentation process gave them the chance to review their work so that it became a record of their learning. The educators valued the documentation as a valuable record of events and a catalyst for conversations. The learning stories created were a testament to the changes in their practices and their renewed focus on play-based inquiries—and their role at the centre of learning and often provided the context to re-launch ideas and inquiries at a later date.

Pedagogies

The STEM Bridge project involved the children and their teachers in many different inquiries during the course of the year. We began to differentiate the variety of new pedagogical 'moves' that the educators and teachers were using and to refer to them collectively as a 'pedagogical repertoire'. These included the pedagogies of:

- Rethinking and creating learning ecologies that embodied the following:
  - Pedagogical conversations
  - Questioning

- Noticing
- Describing, analysing and recording
- Re-launching ideas and inquiries
- Inquiry-based learning in explorations
- Scaffolding learning via a variety of processes
- Knowledge building

• Reflecting on learning with peers, and with pedagogical documentation and learning panels

Pedagogical practices were transformed. The transformation was made possible by preschool educators and teachers across all the sites making time to collaborate, discuss and implement inquiries. The principals and ECLs participated in these design and conversation days and the teachers noted that having this support from their leaders was an important factor in their ability to make dramatic pedagogical shifts. They could justify their new pedagogical approaches, knowing their ideas would be supported.

Narratives of Authentic, Interdisciplinary Inquiries

Inquiry: Learning in the Wirra

At one of the sites there was a patch of native vegetation called the Wirra (Fig. 12.1). Initially, the Reception teacher was reluctant to take her class to explore the Wirra as part of their everyday learning experiences because the administrative processes associated with working outside the classroom were complex. However, when these were considered in a practical way with the support of the principal, it was made possible by viewing the time in the Wirra as simply an extension of the classroom and a rich place of learning. The short walk to the Wirra within the context of the school grounds meant that it was not considered as being 'off site' because it was adjacent to the preschool and school. The area became the focus of shared learning experiences for the preschool and Reception children on a weekly basis from the start of the project and continued on a

Fig. 12.1 The Wirra

regular basis. These shared learning experiences were in contrast with the traditional orientation-focused transition process which occurred at the end of the year when the pre-schoolers joined the Reception class for two visits and did some fun activities. Thus, the concept of transition as orientation was immediately changed to one in which collaboration and shared learning ecologies were foregrounded.

As stated above, the challenge of using outdoor space was noted by all teachers as a cause of concern in their consideration of creating new contexts for learning. They had come from the perspective that it was too difficult to organise and thus avoided it. However, with the new imperatives created by participation in the STEM Bridge project, these shared inquiry times became a focal point for deep learning and inquiry. Teachers noticed, for example, that 'compliant, quiet children' became confident talkers and observers in the context of this outdoor learning. The two groups and their teachers engaged in what they called, 'Wirra

Fig. 12.2 Wonderings about the Wirra

wonderings' (Fig. 12.2), For example, one child noticed the wild orchids growing in the Wirra and suggested that we needed to put markers around them so that visitors would know they were there and would not step on them. Another source of wonder was the different types of plants that were identified as native and 'introduced' soursobs (scientific name: Oxalis pes-caprae)—which ones could they feed to the chickens at the kindergarten?

The preschool educators and teachers noted the following in their journals:

- "We have created time and space in our program for the children to connect with nature within a familiar shared space."
- "We have been carefully considering the types of questions we use in our Wirra wonderings. For example, shifting from: 'What do you know about birds?' to, 'What are your ideas about birds?'"

- "We have a vision of STEM learning in the early years as being playful and beginning in the children's curiosity. We have been exploring how we can plan to build on these experiences back in the centre/classroom"
- "We have a shared vision about why we are doing this – it is because we want our children to connect kindy experiences with school experiences to make their transitions smooth and relevant to them." [Educator/Teacher journals, 2018, Site A]

When the children returned inside, the educators asked them to 'draw what you see'. In one instance they saw butterflies in the Wirra and then came in to watch a short video called 'Austin's butterfly'. The educators also borrowed butterfly collections from the Nature Education centre and the children began to study the details of the butterflies and record them in their drawings, pointing out the various features as they completed the drawing (Fig. 12.3).

Fig. 12.3 Technical drawings

The educators noted that when given time to explore, the children had sustained opportunities to engage in the mathematical and science process skills of listening, observing, comparing, classifying and describing, and documenting their discoveries in the Wirra. In addition, their follow-up work (e.g. conversations discussions and documentation) after spending time in the Wirra became more detailed and their language more sophisticated. Accordingly, the teachers created more opportunities for the children to notice, talk and record their observations from the Wirra.

Inquiry: Dylan and the Mat in the Playground

In another scenario, Dylan, a Year 5 boy from a primary school not participating in the project, moved a striped mat from the cubby house to under a tree near the boundary between the preschool and school (Fig. 12.4).

The teacher asked the children,: "What was he thinking?" Initially the children voiced their suggestions. For example:

Fig. 12.4 Dylan put the mat under the tree—why?

Fig. 12.5 Why did Dylan do this?

> "[H]e moved that mat from the cubby house to the tree. I think he wanted to lie down on the mat to do something." (Fig. 12.5). "It's really strange. I don't know why?"

The children were then asked: "Should we just leave it there?"

A combined community of inquiry gave every child a voice in this decision. They talked through their reasons and then voted with the majority deciding that the mat should be moved because it did not belong under the tree. Piper said:

> "I don't think we should leave the mat under the tree because it needs to be all natured up!" (Fig. 12.6).

Fig. 12.6 All natured up

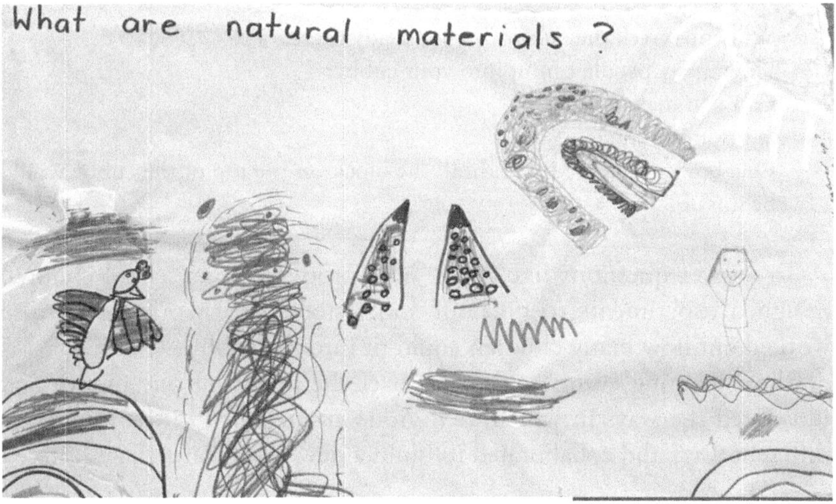

Fig. 12.7 Natural materials

This led the Reception class to think about natural and man-made materials. The question, “What does all natured up look like?” led to drawings that reveal their concepts about these things (Fig. 12.7).

Conversations ensued about what *did* belong under the tree and the children decided it should be a cubby house or a snail playground or something that belonged outside, not like a mat! All children began a design process for a cubby in the spot, drawing on scientific, mathematical, engineering and technology knowledge using systems, design and computational thinking. They used blocks to experiment with possible designs and completed drawings of their plans. In the preschool, another vote occurred to consider the final design for a cubby. They would use loose parts. In the process of building the cubby, the children were engaged in negotiating, using practical building skills, problem solving, rethinking, while continually revisiting and adapting their plans. When the cubby was finalised, they wrote a letter to the Reception class to ask if they thought it was a good design and if they could build it on the shared site.

The Reception children replied with some questions and statements like:

> Will you wreck the cubby that is already there?
> How many people can fit into your cubby?
> Can you make it bigger?
> Is it waterproof?
> One boy commented: We think the block on the top of the cubby would be unsafe?

To answer questions like, "Is it waterproof?" the preschool children designed experiments to find out and tested their theories. They also worked out how many children could fit into the cubby.

These learning events took place over the period of one month and illustrated the ways in which each group participated in the inquiry in different ways and collaborated to build a new cubby on a shared site.

Inquiry: Playing with Shadows

One of the important parts of the project was the use of pedagogical documentation in the form of 'panels' to which both teachers and children contributed (Rintakorpi & Reunamo, 2017). For example, the children's explorations of shadows, light and colour were documented in a panel called 'Shadows' (Fig. 12.8). In the panels, the teachers' summary

The children began to notice their shadows while exploring the outdoor environment. Curious about the position and movement of shadows, they began to question. Together the educators and the children entered into deep conversation about shadows and *how* they move. We listened and actively responded to the children's voices to support their emerging theories. We documented and tested their theories and relaunched ideas and actions to deepen the learning through the inquiry process.

Deepening child-led learning...exploring children's theories

Eric's theory was that the movement of shadows occurred due to the movement of the sun. The educators relaunched this idea through a series of questions that responded to his emerging theories.

How and why did the sun move Eric?

"Cos it's Winter, some days it's Winter.

The earth spins around and the sun shine at different locations that means, different places.

So, how sun moves? Sometimes the sun moves in a day and each day the sun moves."

Why does the sun move each day?

"Maybe because the night, when it moves to a different place and it stops at daytime. Sometimes day it moves. It drives. I think it moves and moves and moves ...it looks like a different place for the sun. I can see sun in a different place. I think it follows the car when you are driving...it follows you."

How does the sun follow you?

Maybe the moon moves and then...the sun goes to the same place. The moon follows my car at night. I think IT moves...moves to places. Maybe...to different countries. I think it always follows you. It stops in different places.

"The sun is going down and the moon is coming up. Follows your car. The sun went to a different window."

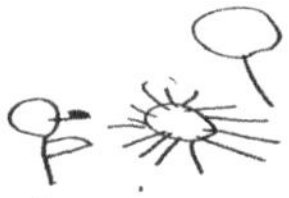

"The moon moves and the sun comes up at the same location" Thomas

It moves, the sun moves...cos...it moves all around and...moves to different places. Zarliah

Fig. 12.8 Shadows

and reflections are evident as well as the children's recorded observations and their theories about making colours and how light works.

The documentation panels were shared as a resource for the other teachers to explore the concepts, supporting teacher collaborative practice. This proved to be a major factor in the transformation of pedagogical practices. One teacher commented, "I did not realise I could do it.... until I saw it .... Then it was so exciting... I had to try it!"

The learning often began when the children noticed and were curious about their own shadows and conversations sprang up between children and with the teachers. Questions were posed and theories discussed. For example, Eric's theory was:

the movement of shadows occurred due to the movement of the sun.

The educators then re-launched this idea through a series of questions that responded to his *emerging* theories, and enabled him to articulate his ideas:

How and why does the sun move Eric?
Why does the sun move each day?
How does the sun follow you?

Other children also participated, drawing and explaining their reasoning about shadows to the educator. These drawings are also included in the panel.

Conclusions

In the STEM Bridge project, preschool educators and Reception teachers at each research site explored the ways in which they could elevate pedagogical conversations around children's voice. The practices of noticing and generating questions formed the basis of investigations, and extending play-based inquiries into the school cosntext was a major feature of the preschool and school collaborations. In rethinking the learning ecologies in which the young children were immersed, the educators reflected on the changes in their own ideas and attitudes to learning over the year.

By the end of the year, the teachers were describing their pedagogical shifts as being 'child led' with teacher guidance and scaffolding supporting the children's deep learning. This had come about because of the collaborative approach to planning that the preschool educator and Reception teachers initiated for the purposes of the STEM Bridge project. For all participants, the teacher's role changed to being more proactive and provocative—challenging each other's thinking in new and dynamic ways. The Reception teachers, in particular, became much more flexible in designing and extending their learning ecologies into the outdoor areas available to them. For Reception teachers, the starting points, or provocations, for explorations were later to be described as 'natural' and achievable at multiple entry and exit points. They recognised that the children were becoming more proactive in their response to provocation-based inquiries. They wanted to be challenged and they definitely wanted to make use of valuable outside learning environments. Additionally, the teachers were using a wider range of pedagogical strategies and more sophisticated questioning techniques. Teachers were extending conversations and re-launching ideas and inquiries at a later date.

The preschool educators highlighted their increased awareness of the value of flexible approaches to the use of space and shared materials, and of building connections between the preschool and school environs, and approaches to pedagogies and learning. They also reflected on the ways in which their different curriculum imperatives could be incorporated into these changing learning ecologies. Finally, they reflected on the shift from a traditional view of transition as being 'school ready' to one in which discussions of pedagogy were elevated to focus on the learner. For the Reception teachers this also involved being challenged to justify their new pedagogical approaches in the context of achieving the mandated aims and outcomes of learning areas in the National Curriculum (ACARA, n.d.) by the Year 1 and Year 2 teacher colleagues. In one instance the Reception teacher smiled as she told us the story of how the Year 1 teacher had said the children were just "playing"... and thought that when they came into Year 1 the following year, they would not have mastered the required content knowledge. The experienced Reception teacher, who had been at the school for seven years, responded by referring to her planning documentation that indicated that the class were in fact studying:

> *Science* as understanding, as a human endeavor and using the Science inquiry skills as they:
>
> Studied that living things had basic needs [ACSSU002]
> Understood that objects are made of materials that have observable properties [ACSSU003]
> (Earth Sciences) Recognised that there were daily and seasonal changes in their lives [ACSSU004]
> Asked questions and described and recorded their findings [ACSHE013]
> Made predictions about familiar objects and events [ACSIS014]
> Were engaged in planning investigations and making observations with all their senses [ACSIS011]
> Engaging in discussions about their observations and reporting their ideas and theories [ACSIS012]
> Communicating and sharing their observations and finding in multimodal formats [ACAIA012]
>
> and
>
> *Mathematics* when using number and place value [ACMN002], comparing and ordering, and making comparisons between collections [ACMN289], and using patterns, measurement and geometry in their investigations. This included using statistics and probability to create and answer questions about their *wonderings and noticings* and generating theories about their world (e.g. why shadows moved their position?)

The most overt changes were evident in the pedagogical practices of the Reception teachers as they incorporated play-based pedagogies and STEM learning ecologies that extended beyond their classrooms.

> *This project has provided me with the opportunity to experience and support learners in a play-based environment.*
>
> *Educators mentor and support each other around possibilities and missed opportunities as we reflect and modify our practices.*

For example…. Feeling supported and having permission to continue with sustained interactions in play spaces. (Reception Teacher, STEM Bridge Project)

Learning Pathways

In this chapter, STEM education has been conceptualised as providing a context for active learning and inquiries that respond to children's natural curiosity about the world in which they live. Preschool educators and Reception teachers purposefully provided opportunities for children to apply critical, creative and design thinking skills as they used their skills to build knowledge, and acquire confident dispositions for learning.

These ways of *doing* pedagogies are prioritised:

- Teachers being engaged in reflective practice and having pedagogical conversations about their work in open and collegial ways across two sectors of education, which traditionally have only casual encounters as the end of a school year approaches.
- Engaging in rich conversations that lead to deep collaborations between all the educators. Here, these went beyond superficial actions and included opportunities for teachers mentoring and coaching each other which broadened each teacher's pedagogical repertoire.
- A cross-sector, whole-system approach which was supported by principals and the Department for Education so that all participants were investing in the desire to transform their pedagogies to support deep learning.
- The importance of building time into the curriculum for all educators so that they were able to design and enact a STEM inquiry approach that was meaningful and relevant to themselves and the children.

In practice, the STEM Bridge project illustrated the following:

- STEM learning ecologies can act as a catalyst for creating communities of practice with a focus on inquiry, active learning and creative thinking.

- The importance of being fluent in asking questions that promote creativity and extend thinking (e.g. What are your ideas about birds?) and sharing these strategies so that parents can build on them at home.
- Creating play-based learning scenarios that encouraged the use of STEM processes (e.g. noticing, observing, questioning, classifying, describing) enabled inquiries that led to knowledge building. In constructing their own knowledge, the children were interacting with curriculum content in a personal way that gave every child an intimate sense of belonging and engagement that ensured deep learning.

References

Aldemir, J., & Kermani, H. (2016). Integrated STEM curriculum: Improving educational outcomes for head start children. *Early Childhood Education and Care, 187*(11), 1694–1705.

Australian Curriculum Assessment and Reporting Authority (ACARA) (n.d.). *F-10 curriculum*. https://www.australiancurriculum.edu.au/f-10-curriculum/

Bennett, N., Wood, L., & Rogers, S. (1997). *Teaching through play: Teachers' thinking and classroom practice*. Buckingham, Philadelphia: Open University Press.

Blackley, S., & Howell, J. (2015). A STEM narrative: 15 years in the making. *Australian Journal of Teacher Education, 40*(7), 102–112. https://doi.org/10.14221/ajte.2015v40n7.8

Brenneman, K., Lange, A., & Nayfeld, I. (2019). Integrating STEM into preschool education; developing a professional development model in diverse settings. *Early Childhood Education Journal, 47*, 15–28.

Carter, A. (2016). Why is there such an emphasis on STEM subjects. *Science Educational News, 65*(3), 11–19.

Cohrssen, C., & Page, J. (2016). Articulating a rights based argument for mathematics teaching and learning in early childhood education. *Australasian Journal of Early Childhood, 41*(3), 104–107.

DEEWR (Department of Education Employment and Workplace Relations). (2009). *Belonging, being and becoming: The early years learning framework for Australia*. Retrieved from http://docs.education.gov.au/system/files/doc/other/belonging_being_and_becoming_the_early_years_learning_framework_for_australia.pdf

Early Childhood STEM Working Group. (2017). *Early STEM matters*. Providing high quality STEM experiences for all young learners. Retrieved 25 August 2018, from http://d3lwefg3pyezlb.cloudfront.net/docs/Early_STEM_Matters_FINAL.pdf

French, L. (2004). Science at the center of a Coherent, integrated early childhood curriculum. *Early Childhood Research Quarterly, 19*, 138–149.

Fusaro, M. S., & Smith, M. C. (2018). Preschoolers' inquisitiveness and science-relevant problem solving. *Early Childhood Research Quarterly, 42*, 119–127.

Hofer, K. G., Farran, D. C., & Cummings, T. P. (2013). Preschool children's math-related behaviors mediate curriculum effects of math achievement gains. *Early Childhood Research Quarterly, 28*, 487–495.

John, M. S., Sibuma, B., Wunnava, S., Anggoro, F., & Dubosarsky, M. (2018). An iterative participatory approach to developing an early childhood problem based STEM curriculum. *European Journal of STEM Education, 3*(3), 2468–4368.

Katz, L. (2010). STEM in the early years. *Early Childhood Research and Practice*, (Fall), 1–7.

Kelley, T. R., & Knowles, J.G. (2016). A Conceptual framework for integrated STEM education. *International Journal of STEM Education, 3*(11). https://doi.org/10.1186/s40594-016-0046-2

Kemmis, S., & McTaggert, R. (2007). Retrieved 19 August 2019 from http://citeseerx.ist.psu.edu/viewdoc/download?doi=10.1.1.473.4759&rep=rep1&type=pdf

National Innovation in Schools (NISA). (2015a). *Welcome to the ideas boom*. Agenda report. https://www.industry.gov.au/sites/default/files/July%202018/document/pdf/national-innovation-and-science-agenda-report.pdf?acsf_files_redirect

National Innovation in Schools (NISA). (2015b). *National STEM education schools strategy*. Canberra, ACT: Education Council for Australia. http://www.educationcouncil.edu.au/site/DefaultSite/filesystem/documents/National%20STEM%20School%20Education%20Strategy.pdf

Partnerships for the 21st Century. (2008). 21st Century skills, education & competitiveness: A resource and policy guide. Retrieved from Washington, DC: https://files.eric.ed.gov/fulltext/ED519337.pdf

Rintakorpi, K., & Reunamo, J. (2017). Pedagogical documentation and its relation to every activities in early years. *Early Childhood Education and Care, 187*(11), 1611–1622.

Schweinhart, L. J. (2007). Outcomes of the High/ Scope Perry Preschool Study and Michigan School Readiness Program. Retrieved from https://www.

researchgate.net/publication/254108264_Outcomes_of_the_HighScope_Perry_Preschool_Study_and_Michigan_School_Readiness_Program.

Strauss, A., & Corbin, J. (1990). *Basics of qualitative research: Grounded theory procedures and techniques*. Newbury Park, CA: Sage.

Trilling, B., & Fadel, C. (2009). 21st Century skills: Learning for life in our times. San Francisco: Jossey-Bass.

Yelland, N. J. (2011). Reconceptualising play and learning in the lives of children. *Australasian Journal of Early Childhood, 36*(2), 4–12.

Yelland, N. J. (2018). A pedagogy of multiliteracies: young children and multimodal learning with tablets. *British Journal of Educational Technology, 49*(5), 847–858.

13

STEM or STEAM or STREAM? Integrated or Interdisciplinary?

Douglas H. Clements and Julie Sarama

Introduction

This book exemplifies a growing interest in STEM (Science, Technology, Engineering, and Mathematics) education in the early years (Committee on STEM Education, 2018; McClure et al., 2017; Sarama et al., 2018; Sarama, Brenneman, Clements, Duke, & Hemmeter, 2017). We discuss two tendencies in this movement. The first is the addition of one or more domains, resulting in the acronym STEAM or STREAM. The second is the notion that the best educational approach is to fully integrate these and all other domains. We present arguments and evidence that these tendencies appear positive, but may inadvertently negatively impact the critical increase of STEM in early education. We present an alternative approach that maintains the positive aspects of these tendencies while avoiding the possible negatives.

D. H. Clements (✉) • J. Sarama
University of Denver, Denver, CO, USA
e-mail: Douglas.Clements@du.edu; Julie.Sarama@du.edu

C. Cohrssen, S. Garvis (eds.), *Embedding STEAM in Early Childhood Education and Care*, https://doi.org/10.1007/978-3-030-65624-9_13

STEM or STEAM or STREAM or…?

Although STEM is the original and prominent acronym, many, including Head Start in the USA, prefer the acronym STEAM, to emphasize the inclusion of the arts. Although the importance of arts education cannot be denied, we believe that extending the acronym is unwise, as it both loosens the coherence of the STEM domains and excludes other domains that are as, or more, important to the construction of rich interdisciplinary educational experiences for young children. We discuss three reasons for this position: (a) STEM is a serious movement across ages and countries with which we should align, (b) adding an “A” to STEM and leaving other domains out is not justifiable, and (c) adding the arts or other domains weakens the coherence of the STEM domains, which differ from others. We discuss each of these in turn.

The STEM Movement, Birth Through University

STEM is a serious movement across ages and countries with which we should align. One of us is a member of the national STEM Education Advisory Panel, and the acronym and activities are always STEM, not something else. One might argue that all organizations should change the acronym, but that is extremely unlikely. Instead, we early educators would increase the unfortunate disconnect between us and others. Again, one might respond that the other domains simply cannot be ignored! The next two reasons address that issue.

What Domains Should Be Included

Adding an “A” to STEM or an “R” and “A” (R for Reading, Furman, 2017) is understandable but not justifiable. To begin, if one is to add the arts, why not music (the “arts” often implies visual forms, although this is a limited interpretation), among other forms of expression? Specific gross (athletics) and fine motor (Grissmer, Grimm, Aiyer, Murrah, & Steele, 2010) domains should not be ignored, either.

Perhaps more important, why not language and literacy and social-emotional development, two clearly critical domains? Arguably (and supported by research evidence), done well, STEM contributes more to authentic language and literacy and social-emotional development than it does to the arts. Exaggerating to make a point, the unpronounceable initialism would then stand as STREAMMLLSEFMGM (STREAM + Music, Language, Literacy, Social, Emotional, Fine Motor, Gross Motor).

A devil's advocate might say, "We have enough literacy! Enough social-emotional! We don't need to add those!" However, physical education is often underemphasized, as are music, dance, and other forms (music is often "done"—but not well!). Further, the STEM movement is simply the wrong place for those with justifiable desires to enrich the curriculum to "add on" their own preferences. One good reason for this is that STEM *is* a coherent domain, different from the others—our third reason.

The Coherence of STEM

Perhaps the most important reason to maintain STEM is that adding the arts, reading, or other domains weakens the subject matter content, pedagogical, and epistemological *coherence* of STEM, which differ substantially from the other domains. Considering content, science, technology, and engineering are a tight group, with science providing the concepts and processes used in creating and refining new technologies and engineering advances and explaining *why* some unexpected feats of engineering work. Similarly, technology and engineering put science to work, ideally for the good of humanity and the planet.

What is the role of mathematics? Mathematics remains the queen of all science (and technology and engineering) both in its realization in the world and educationally. Almost every STEM advance and project is expressed in the language of mathematics.[1] Educationally, mathematical development is central. What mathematics children know when they *enter* kindergarten predicts their mathematics achievement for years to

---

[1] For example: "Mathematics is the alphabet with which God has written the universe" (Galileo). "How is it possible that mathematics, a product of human thought that is independent of experience, fits so excellently the objects of reality?" (Einstein).

come (Duncan et al., 2007). Mathematics also predicts later success in *reading* (Duncan et al., 2007; Duncan & Magnuson, 2011), so *mathematics appears to be a core component of cognition.* Further, knowledge of mathematics in the early years is the best predictor of graduating high school (McCoy et al., 2017; Watts, Duncan, Siegler, & Davis-Kean, 2014). Finally, number and arithmetic knowledge at age 7 predicts socioeconomic status at age 42 (even controlling for all other variables; Ritchie & Bates, 2013).

Mathematics is also important in more advanced education. The more mathematics courses high school students take predicts their university performance not just in mathematics, but in biology, chemistry, and physics—with a strength that matches or excels the amount of those subjects students study (Sadler & Tai, 2007).

Thus, both in subject matter content and in educational contexts, there is a tight, coherent linkage among the STEM domains. This does not imply that there are no connections to other domains. High-quality experiences in early science and mathematics, for example, have been connected to later development of language, reading, and executive function competencies (e.g., Duncan et al., 2007; Institute of Medicine (IOM) and National Research Council (NRC), 2015), even in rigorous causal studies (Clements, Sarama, Layzer, Unlu, & Fesler, 2020; Sarama, Lange, Clements, & Wolfe, 2012). Further, these relationships are bidirectional (Clements, Sarama, & Germeroth, 2016; Purpura, Hume, Sims, & Lonigan, 2011). However, these are smaller and looser connections that emerge from one domain *supporting* the other (distinct) domain educationally.

STEM differs markedly from the arts in epistemology, or "how you know you know something" in a domain. So, just shoving STEM together with the arts to make a point about foci confuses important issues that we believe should be clarified. Let us begin this discussion by noting that even within the STEM field there is one important difference in how you find "truth." Validity in mathematics comes from logic, reasoning, and proof—it's "in your head." Validity in STE stems from scientific testing of ideas and theories in the world. Even preschoolers implicitly learn these knowledge foundations. One four-year-old said to another, "You don't have to ask the teacher. Triangles have three sides connected. This one is really skinny, but it's *got* that. It has to be a triangle!" In contrast, in the same classroom, the teacher recorded a long discussion ending with, "We don't know if this design is the best. We need to test it." Despite this difference in knowledge foundations between STE and M, however, the tight linkages of STEM make these domains remarkably complementary and dependent on one another.

In contrast, arts and literature are a distinctly different way of *understanding* humanity and our world. The nature of knowledge and "truth" is distinct. Jerome Bruner puts it this way.

> *There are two modes of cognitive functioning, two modes of thought, each providing distinctive ways of ordering experience, of constructing reality. The two (though complementary) are irreducible to one another. Efforts to reduce one mode to the other or to ignore one at the expense of the other inevitably fail to capture the rich diversity of thought.*
>
> *Each of the ways of knowing, moreover, has operating principles of its own and its own criteria of well-formedness. They differ radically in their procedures for verification. A good story and a well-formed argument are different natural kinds. Both can be used as means for convincing another. Yet what they convince of is fundamentally different: arguments convince one of their truth, stories of their lifelikeness. The one verifies by eventual appeal to procedures for establishing formal and empirical proof. The other establishes not truth but verisimilitude. It has been claimed that the one is a refinement of or an abstraction from the other. But this must be either false or true only in the most unenlightening way.* (Bruner, 1986, p. 11)

The first, Bruner calls the "logico-scientific," and it includes all of STEM. The second, the narrative/artistic way of knowing, includes reasoning and experience, of course, but through literature and the arts, interpreting the meaning of existence, beauty, and the "verisimilitudes of intention" (p. 11). As Einstein said, "Not everything that can be counted, counts; and not everything that counts, can be counted."

Thus, we return to a critical point: *Yes,* we want arts and literature, social and emotional development, all seamlessly connected to STEM. For example, we know science experiences are related to children's vocabulary growth (French, 2004) and use of more complex grammatical structures, such as causal connectives (Peterson & French, 2008). However, expanding STEM to include these other domains muddies the conceptual waters and expands STEM's nature to incoherence. How then should we connect them? That is the topic of the next section, addressing the second tendency in early childhood STEM.

Integrated or Interdisciplinary?

Given the connections, it is reasonable to claim that curricula and pedagogical approaches should *fully integrate* all aspects of STEM and other domains. That is, every planned or emergent experience should be guided to include all valued domains. However, the history of such educational efforts and research evaluations raise concerns about such complete integration.

Problems with the Fully Integrated Approach

Research comparing various types of curricula do not support such full integration. For example, reviews of fully integrated curricula (e.g., activities that involve all subject matter areas) in preschool and higher reveal little evidence that they are superior to traditional structures and that there are challenges in implementing such curricula (Czerniak, Weber Jr., Sandmann, & Ahern, 1999; George, 1996; Preschool Curriculum Evaluation Research Consortium, 2008).

Why *might* this be so? One hint emerged from half-century-old integrated science and mathematics National Science Foundation (NSF)-funded projects such as USMES (Ellis, 2004). USMES is an acronym for Unified Science and Mathematics for Elementary Schools, although it was informally renamed by some Unified Science and Mathematics and English for Schools due to the large amount of language and literacy included. One of us worked with local fifth and sixth grades implementing units. The integration of all these domains was clear. The problem was that the mathematics almost never surpassed adding and subtracting two-digit numbers. This taught nothing *in mathematics* to these intermediate-grade students. They needed daily work with division, fractions, advanced measurement and geometry, and so forth—and the only mathematics they did for months was second grade level. The project took precedence over the needs of the subject. One sees similar examples in early childhood. For example, planting seeds in Spring is good for science in pre-K or Kindergarten. Counting those that germinated for each student is *not* good mathematics for those children.[2]

Creating the Interdisciplinary Approach

For this reason and based on our belief that each domain requires some unique strategies for teaching and learning, we created an *interdisciplinary approach*, in which rich connections are made between domains, but each retains its core conceptual, procedural, and knowledge foundation (epistemological) structures. That is, two or more domains would always (albeit only) be integrated when that combination is both consistent and complementary with those structures for *each* domain. Thus, for example, children gain exposure to prerequisite mathematics skills in an appropriate sequence and science inquiry is designed to promote a deep understanding of conceptual content and science processes. On the other hand, when connections *are* drawn between mathematics and science, they are genuine and detailed, with their impact undiluted by less fruitful attempts at integration.

To realize the potential of this approach, we created an interdisciplinary curriculum for science/engineering, mathematics, literacy, and social-emotional learning called "Connect4Learning" (C4L, Sarama, Brenneman, Clements, Duke, & Hemmeter, 2016). The "4" in C4L refers to the fact that most children in pre-K, the setting we are targeting, are 4 years old and to the 4 domains we emphasize. And, of course, we use the homophone ("four"/"for") to emphasize that we connect the domains *for* learning. That is, we provide supports to teachers and children to make connections within and among the domains to support the learning and development of the "whole child." Thus, we believe that it is possible to provide high-quality learning experiences for young children

[2] Especially for our own children, except for possibly their concept of zero! They routinely brought home cups containing only very wet mud.

across *all* critical domains—not only in the language and literacy and social-emotional domains—and that the fundamental academic domains of STEM provide rich content on which to build these learning experiences. We integrate them whenever it is advantageous to *each of the domains so connected* but do not force integration. Further, "we" means all the experts we collaborated with us on C4L: Dr. Kimberley Brenneman in science and engineering, Dr. Nell Duke in language and literacy, and Dr. Mary Louise Hemmeter in social and emotional development.

Working together, these five authors, along with many graduate research assistants, especially Lindsay N. Giroux and Lynne M. Watanabe, created a series of units, integrating instructional activities when appropriate—*if and only if* such integration represents a happy alignment in which the cognitive activity serves children's development along learning trajectories in two or more core domains. Thus, we determined the best way to structure the complex interaction among the domains. One strategy was to begin with mathematics for which there is a research-based developmental sequence of core concepts and core process skills (Clements & Sarama, 2014; Sarama & Clements, 2009) and then to *determine connections to science.* Based on this analysis, the science units could often be sequenced to maximize opportunities for integration, allowing these units to influence the placement and sequence of the relatively independent (e.g., geometry vs. number) mathematics learning trajectories.

Language and literacy competencies were structured into these mathematics-and-science units, informed by the broader learning trajectories of language and literacy (e.g., phonemic awareness and alphabet recognition to early graphophonemic analysis). We employed pedagogical strategies for implementing these instructional activities in ways known to foster social-emotional development and self-regulation. Further, we used social contexts and practices to provide a context for providing instruction across domains. For example, Think-Pair-Share is used during read-aloud and cooperative learning activities are used to teach content from other domains.

We all recognized the disadvantages of this interdisciplinary approach, especially that the careful curricular planning and integration it requires are quite difficult. Extending our previous example, we wanted to do the garden unit in the Spring but agreed that counting the number of seeds each child germinates did not fit our mathematical learning trajectories, which were far beyond such simple counting. Therefore, this case, the science topic determined the sequence and we planned that some arithmetic operations would be involved. Moreover, we focused most of the mathematics activities on geometric shape composition (e.g., making pictures of flowers by composing shapes to make compound shapes), which were developed at a similar time of year in the original learning trajectories (e.g., Clements & Sarama, 2004, 2014; Sarama & Clements, 2009). This is a main reason we created the C4L curriculum, to spend years wrestling with these difficulties to provide examples of research-based interdisciplinary educational units.

An Example Unit

One C4L unit focuses on how structures and tools work and how to make and do things (Clements, Sarama, Brenneman, Duke, & Hemmeter, 2020; Sarama et al., 2016). As an example of rich integration among four domains, teachers read Albert's Alphabet, by Leslie Tryon, in which Albert the duck reads a letter from his supervisor asking him to make all 26 letters throughout their park. Children describe (many pages have no text) and discuss how Albert used materials and forms. They reason about *structure* and function—straight, rigid lumber works build letters like A and E, but B, C, and D need curved pipe or flexible materials. Later, children worked together to choose materials to make their own letters such as cutting paper straws to make line segments for an F, but flexible chenille sticks for the curves of a C. Children found yarn to be flexible enough, but it did not hold its shape. Science includes physics and the identification of attributes of materials, technology, and engineering in the choice and manipulation of materials and recognition of form and function, and mathematics in the geometric relations (one child: "A has two straight lines the same length, one slanted this way, one slanted that way, and a shorter straight line between them. That makes a triangle at the top!"). The literacy connections are clear, from books to letter naming (and "writing"), and positive social interactions are engendered as children solved problems cooperatively and shared solutions.

The themes continue with a read-aloud of Spoon, by Amy Krause Rosenthal. The main character, Spoon, wishes he were more like his friends Fork, Knife, and Chopsticks because they can twirl spaghetti (Fork) and cut food (Knife). Children discussed form and function questions such as "Why is Fork better than Spoon for twirling spaghetti?" and "What parts does Fork have that Spoon doesn't have?" Later in the read aloud, when Spoon's mother helps him to recognize his own special talents, the children discussed questions such as "What is special about a spoon that makes it a good tool for eating soup?" In a discussion of this literature, children discuss not only structure and function, but also social and emotional learning themes. Later, they engaged in a chopstick challenge, attempting to pick up beads and pom-poms with chopsticks, and made connections about how an object's form, material, and attributes affect its function. Note that they saw that technology is not just computers—it includes tools like chopsticks, knives and scissors.

Thinking about the design and function of everyday kitchen tools continued with a visit from Conrad the Confused Chameleon. Conrad couldn't find a bowl for his breakfast cereal. Children help him understand why, despite its shape, a colander was not a good substitute and why a flat plate—despite having no holes—wasn't either! In small-group learning experiences and related centers, the children had more opportunities to explore the structures and functions of common tools (technologies).

What experiences were *less* integrated but just right for the children? Activities on calming strategies and solving social problem were clearly in the social-emotional domain. Although they included language development, STEM is not forced into these concentrated activities. In mathematics, playful fast focus activities help children (a) recognize small quantities—like three beads—quickly without counting (known as "subitizing") and (b) construct understanding of the connection between counting and simple addition and subtraction (add one, the result is the next counting number). Similarly, the language arts are emphasized in "fast focus" lessons on P (name and linking to its sound) and on the features and use of a how-to text (materials, steps—preparing for the final project as we'll see). The children were strongly invested in following the procedure for making their snack, ants on a log.

All these build the foundation for the unit project, in which children imagine, design, and create toys. In small groups, children worked as engineers to first determine the best ball to use (various options were available) and then figure out (by testing their ideas) what size hole they'd need to create to make a good scoop to catch their ball. They eventually produce a procedural text on how to make a ball and scoop using a research-supported instructional technique called *interactive writing* (Hall et al. 2014). That is, the teacher is the primary writer; however, the children and the teacher compose the text together, with the children contributing some of the writing.

Creating and sharing the toys and informational texts complete the curricular unit, but the preschoolers also keep duplicate toys they make. They use these toys (and many others) in dramatic play, where they pretended to buy and sell toys in their own toy store. They made signs, lists, and price tags and then took the roles of cashier, bagger, wrapper, and shipper. So, once again, we see carefully planned, fruitful integration.

Final Words: What We Are and Are Not Saying

We argue that extending the STEM acronym is unwise, as it simultaneously weakens the coherence of the STEM domains and excludes other domains that are critical to the construction of rich interdisciplinary educational experiences for young children. We provided three reasons for this position.

- STEM is a serious movement across ages and countries with which we should align.
- Adding other subjects to STEM and leaving other domains out is not justifiable.
- Adding other domains weakens the content, pedagogical, and knowledge foundation coherence of STEM.

We are not saying STEM is more important than or unrelated to these other domains. Young children need high-quality experiences with them all.

We further argue that a fully integrated approach to early childhood education, within which all experiences are guided to include all valued domains is also unwise. Instead, we define an interdisciplinary approach as one in which rich connections are made between domains, but each retains its core conceptual, procedural, and epistemological structures.

Thus, we are *not* saying domains should be taught separately—quite the opposite. We celebrate integration, but *only as it serves the needs of each domain*. Thus, we believe educational planning, simultaneously from balanced and interactive integrated and separate disciplinary perspectives, provides the "best of all worlds."

The interdisciplinary approach promoted here suggests the following guidelines (Clements, Sarama, Brenneman, et al., 2020).

- Don't underestimate what children can do in each domain.
- Incorporate research-supported practices: Specific techniques inside and outside of STEM, such as providing practice with subitizing and interactive writing, can be embedded within and contribute to the unit project's purpose.
- Establish a real-world purpose for children's STEM projects.
- Consider the role of each domain in the project: It may be easier to see where the science and mathematics come in but be sure to consider the technology and engineering as well as literacy, music, the arts, gross and fine motor, and so on.
- Take an interdisciplinary approach. Look for all possible connections between domains, but avoid forcing integration.
- Attend to all of the science disciplines: Too often, science education with young children focuses on biology (such as learning about animals and habitats) to the exclusion of earth science, engineering, and physical science.

Acknowledgment This work was supported in part by the National Science Foundation under Grant No. DRL-1020118. Any opinions, findings, and conclusions or recommendations expressed in this material are those of the authors and do not necessarily reflect the views of the NSF.

References

Bruner, J. (1986). *Actual minds, possible worlds*. Cambridge, MA: Harvard University Press.

Clements, D. H., & Sarama, J. (2004). Learning trajectories in mathematics education. *Mathematical Thinking and Learning, 6*, 81–89. https://doi.org/10.1207/s15327833mtl0602_1

Clements, D. H., & Sarama, J. (2014). *Learning and teaching early math: The learning trajectories approach* (2nd ed.). New York: Routledge.

Clements, D. H., Sarama, J., Brenneman, K., Duke, N. K., & Hemmeter, M. L. (2020). STREAM education at work—No, at play! A toy-making unit. *YC Young Children, 75*(2), 36–43.

Clements, D. H., Sarama, J., & Germeroth, C. (2016). Learning executive function and early mathematics: Directions of causal relations. *Early Childhood Research Quarterly, 36*(3), 79–90. https://doi.org/10.1016/j.ecresq.2015.12.009

Clements, D. H., Sarama, J., Layzer, C., Unlu, F., & Fesler, L. (2020). Effects on mathematics and executive function of a mathematics and play intervention versus mathematics alone. *Journal for Research in Mathematics Education, 51*(3), 301–333. https://doi.org/10.5951/jresemtheduc-2019-0069

Committee on STEM Education. (2018). *Charting a course for success: America's strategy for STEM education*. Washington, DC: Office of Science and Technology Policy. Retrieved from https://www.whitehouse.gov/wp-content/uploads/2018/12/STEM-Education-Strategic-Plan-2018.pdf

Czerniak, C. M., Weber Jr., W. B., Sandmann, A., & Ahern, J. (1999). A literature review of science and mathematics integration. *School Science and Mathematics, 99*(8), 421–430.

Duncan, G. J., Dowsett, C. J., Claessens, A., Magnuson, K., Huston, A. C., Klebanov, P., et al. (2007). School readiness and later achievement. *Developmental Psychology, 43*(6), 1428–1446. https://doi.org/10.1037/0012-1649.43.6.1428

Duncan, G. J., & Magnuson, K. (2011). The nature and impact of early achievement skills, attention skills, and behavior problems. In G. J. Duncan & R. Murnane (Eds.), *Whither opportunity? Rising inequality and the uncertain life chances of low-income children* (pp. 47–70). New York: Sage.

Ellis, A. K. (2004). *Exemplars of curriculum theory*. New York: Routledge.

French, L. (2004). Science as the center of a coherent, integrated early childhood curriculum. *Early Childhood Research Quarterly, 19*(1), 138–149. https://doi.org/10.1016/j.ecresq.2004.01.004

Furman, R. L. (2017). *STEM needs to be updated to STREAM*. Retrieved from website: https://www.huffpost.com/entry/stem-needs-updated-to-str_b_5461814?guccounter=1

George, P. S. (1996). The integrated curriculum: A reality check. *Middle School Journal, 28*(1), 12–19.

Grissmer, D., Grimm, K. J., Aiyer, S. M., Murrah, W. M., & Steele, J. S. (2010). Fine motor skills and early comprehension of the world: Two new school readiness indicators. *Developmental Psychology, 46*(5), 1008–1017. https://doi.org/10.1037/a0020104 and https://doi.org/10.1037/a0020104.supp (Supplemental)

Hall, A. H., Toland, M. D., Grisham-Brown, J., & Graham, S. (2014). Exploring interactive writing as an effective practice for increasing Head Start students' alphabet knowledge skills. *Early Childhood Education Journal, 42*, 423–430.

Institute of Medicine (IOM) and National Research Council (NRC). (2015). *Transforming the workforce for children birth through age 8: A unifying foundation*. Washington, DC: National Academy Press.

McClure, E. R., Guernsey, L., Clements, D. H., Bales, S. N., Nichols, J., Kendall-Taylor, N., et al. (2017). *STEM starts early: Grounding science, technology, engineering, and math education in early childhood*. New York: The Joan Ganz Cooney Center at Sesame Workshop.

McCoy, D. C., Yoshikawa, H., Ziol-Guest, K. M., Duncan, G. J., Schindler, H. S., Magnuson, K., et al. (2017). Impacts of early childhood education on medium- and long-term educational outcomes. *Educational Researcher, 46*(8), 474–487. https://doi.org/10.3102/0013189x17737739

Peterson, S. M., & French, L. (2008). Supporting young children's explanations through inquiry science in preschool. *Early Childhood Research Quarterly, 23*, 395–408.

Preschool Curriculum Evaluation Research Consortium. (2008). *Effects of preschool curriculum programs on school readiness (NCER 2008–2009)*. Retrieved from Government Printing Office website: http://ncer.ed.gov

Purpura, D. J., Hume, L. E., Sims, D. M., & Lonigan, C. J. (2011). Early literacy and early numeracy: The value of including early literacy skills in the prediction of numeracy development. *Journal of Experimental Child Psychology, 110*, 647–658. https://doi.org/10.1016/j.jecp.2011.07.004

Ritchie, S. J., & Bates, T. C. (2013). Enduring links from childhood mathematics and reading achievement to adult socioeconomic status. *Psychological Science, 24*, 1301–1308. https://doi.org/10.1177/0956797612466268

Sadler, P. M., & Tai, R. H. (2007). The two high-school pillars supporting college science. *Science, 317*(27), 457–458. https://doi.org/10.1126/science.1144214

Sarama, J., Brenneman, K., Clements, D. H., Duke, N. K., & Hemmeter, M. L. (2016). *Connect4Learning: The Pre-K curriculum*. Lewisville, NC: Connect4Learning.

Sarama, J., Brenneman, K., Clements, D. H., Duke, N. K., & Hemmeter, M. L. (2017). Interdisciplinary teaching across multiple domains: The C4L (Connect4Learning) curriculum. In L. B. Bailey (Ed.), *Implementing the common core state standards across the early childhood curriculum* (pp. 1–53). New York: Routledge.

Sarama, J., & Clements, D. H. (2009). *Early childhood mathematics education research: Learning trajectories for young children*. New York: Routledge.

Sarama, J., Clements, D. H., Nielsen, N., Blanton, M., Romance, N., Hoover, M., … McCulloch, C. (2018). *Considerations for STEM education from PreK through grade 3*. Retrieved from Education Development Center, Inc. website: http://cadrek12.org/resources/considerations-stem-education-prek-through-grade-3

Sarama, J., Lange, A., Clements, D. H., & Wolfe, C. B. (2012). The impacts of an early mathematics curriculum on emerging literacy and language. *Early Childhood Research Quarterly, 27*(3), 489–502. https://doi.org/10.1016/j.ecresq.2011.12.002

Watts, T. W., Duncan, G. J., Siegler, R. S., & Davis-Kean, P. E. (2014). What's past is prologue: Relations between early mathematics knowledge and high school achievement. *Educational Researcher.* https://doi.org/10.3102/0013189X14553660

Index

C. Cohrssen, S. Garvis (eds.), *Embedding STEAM in Early Childhood Education and Care*, https://doi.org/10.1007/978-3-030-65624-9

N

O

P

The manufacturer's authorised representative in the EU is Springer Nature Customer Service Centre GmbH, Europaplatz 3, 69115 Heidelberg, Germany. If you have any concerns regarding our products, please contact ProductSafety@springernature.com

Printed and bound by CPI Group (UK) Ltd, Croydon, CR0 4YY

30/09/2026

02225460-0008